Memristor Emulator Circuits

Abdullah G. Alharbi • Masud H. Chowdhury

Memristor Emulator Circuits

Abdullah G. Alharbi
Department of Electrical Engineering
Jouf University
Sakaka, Saudi Arabia

Masud H. Chowdhury
Department of Computer Science Electrical
Engineering
University of Missouri–Kansas City
Kansas City, MO, USA

ISBN 978-3-030-51884-4   ISBN 978-3-030-51882-0   (eBook)
https://doi.org/10.1007/978-3-030-51882-0

This Springer imprint is published by the registered company Springer Nature Switzerland AG
The registered company address is: Gewerbestrasse 11, 6330 Cham, Switzerland

Dedicated to our beloved parents, spouses, and children.

Preface

About the Subject

Memristor has recently been recognized as a new and the fourth passive element (along with the resistor, capacitor, and inductor) that establishes the direct relationship between electrical charge (q) and magnetic flux (ϕ). A memristor is perceived as a two-terminal passive element that shows a nonlinear hysteresis behavior for a specific frequency range. Memristor technology is drawing widespread attention nowadays due to its potential applications in conventional memory and logic devices, neuromorphic computing, beyond-binary memory and logic technologies, and many other digital, analog, and mixed-signal applications like modulation systems and oscillator circuits. However, as of now, there is no single device available in the market that can genuinely exhibit memristive behavior for a specific frequency range. In the absence of real and physically fabricated memristors, researchers are still relying on memristor emulators to understand the fundamental concepts, investigate the perceived behaviors, and explore potential applications of the memristor.

There has been a surge of interest in performing different theoretical and experimental works on memristor, considering the enormous potentials of this unique perceived device. However, since there is no practical memristive device available to integrate with any on-chip application, numerous circuit-based emulators and macro-models are being developed to investigate and understand the properties and potential applications of the memristor. These circuit-based models or emulator circuits can mimic some of the perceived behaviors of the memristor. Most of the available emulator circuits and macro-models are based on the TiO_2-based memristor proposed by the Hewlett Packard (HP) Lab in 2008. Although the HP model helped researchers understand and analyze many properties of the memristor, it was later acknowledged that the HP model failed to provide the actual nonlinear behaviors of the memristor. Therefore, as improvements over the HP model, more precise memristor models have recently been proposed to mimic the nonlinear behaviors of both voltage-controlled and current-controlled memristors. These are the Simmons Tunneling Barrier Model (STBM), the ThrEshold Adaptive Memristor (TEAM) Model, the Generalized Memristor Model (GMM), and the Voltage ThrEshold Adaptive Memristor (VTEAM). However, to the best of our knowledge, there is no efficient emulator circuit to mimic the nonlinear electrical behavior of the memristors defined by the STBM, TEAM, GMM, and VTEAM models.

About the Book

This book has been written to address a critical need of the R&D community involved in conducting research on memristors, which is an emerging area in the micro- and nanoelectronics domain. There have been many efforts to develop emulator circuits to study the memristor. However, existing emulator circuits can only emulate the linear electrical response of a memristor because these emulators are based on the ideal and linear HP memristor model, which does not fit the anticipated nonlinear behavior of a real memristor. In this book, a set of generic and practical memristor emulator circuits to analyze and understand the nonlinear behavior of current-controlled and voltage-controlled memristor models have been presented. The proposed emulators can mimic the behavior of well-known nonlinear memristor models like the Simmons Tunneling Barrier Model (STBM), the ThrEshold Adaptive Memristor (TEAM) Model, and the Voltage ThrEshold Adaptive Memristor (VTEAM) model in addition to the simple HP model. These memristor emulator circuits can be built in the laboratory using inexpensive off-the-shelf discrete circuit components.

It is important to note that memristor is perceived as a two-terminal device. Therefore, it may appear in a circuit both in floating and grounded states. The proposed emulator circuit development techniques can be configured for both floating and grounded memristor models. In addition to mathematical modeling and analysis of the proposed emulators, this book provides SPICE simulation and experimental results. The analytical observations and experimental results show that the proposed circuits can mimic the nonlinear behavior of real memristors for a specific frequency range.

The proposed emulators have been used to verify some applications like the Wein Bridge Oscillator. Both series and parallel connectivity of the proposed emulators have been studied. A comparison with the existing emulators is presented to highlight the advantages of the proposed emulators.

Organization of the Book

Chapter 1 presents the basic definition, fingerprints or defining characteristics, and history of the memristor. Also, a brief overview of the memristive device implementation efforts is introduced in this chapter.

Chapter 2 provides a comprehensive review of the existing mathematical models, emulator circuits, and macro-models of current-controlled and voltage-controlled memristors.

Chapter 3 presents a new emulator circuit development approach for current-controlled memristors. This chapter includes circuit implementation and analysis of the proposed emulators for the Simmons Tunneling Barrier Model (STBM) and the ThrEshold Adaptive Memristor (TEAM) model. The application of the proposed emulator in a Wien Bridge oscillator is also demonstrated.

Chapter 4 shows mathematical analysis, SPICE simulations, and implementation of three simple emulator circuits for the current-controlled memristor using second-generation current conveyor (CCII), exponential amplifier, and several passive elements.

Chapter 5 presents generic and practical emulator design techniques for the voltage-controlled models. Circuit implementation and experimental results are also provided.

Chapter 6 presents a set of novel techniques to develop emulators for grounded and floating flux-controlled memristors for analog applications.

Chapter 7 summarizes and concludes this book with a brief overview of future research directions.

Sakaka, Saudi Arabia   Abdullah G. Alharbi
Kansas City, MO, USA   Masud H. Chowdhury

Acknowledgments

The completion of this book would not have been possible without the assistance and encouragement of many people. We appreciate our group members (Zarin Tasnim Sandhie, Farid Uddin Ahmed, Moqbull Hossen, Azzedin EsSakhi, Emeshaw Ashenafi, Abdul Hamid Bin Yousuf, Munem Hossain, Marouf Khan, Nahid Hossain, Mahmood Uddin Mohammed, Lohith Kumar Vemula, and Muhammad Sana Ullah) at the Micro and Nano Electronics Laboratory (MNEL) in the Department of Computer Science Electrical Engineering (CSEE), School of Computing and Engineering (SCE), University of Missouri Kansas City (UMKC), for their cooperation. We also like to thank two of our collaborators, Dr. Zainulabideen Khalifa and Dr. Mohammed Fouda, for their help.

Special thanks to SCE Dean, Kevin Truman, and CSEE Department Chair, Ghulam Chaudhry, for their guidance and continuous support to pursue scholarly work. We like to acknowledge ECE Lab Coordinator, Kevin Kirkpatrick, for giving us access to the testing and measurement equipment in various labs to conduct our experimental work. We appreciate our IT support person, Dave Hanna, who helped us resolve issues related to CAD tools, simulation software, and computing resources used for this research project. We also like to thank Nan Lorenz and Prerana Samant of the CSEE Department at UMKC for their time and effort in proofreading and editing this book. Some parts of the research conducted for this book have been supported by the scholarships provided by the School of Graduate Studies at UMKC and the Saudi Arabian Cultural Mission (SACM) in the USA.

Most importantly, we like to thank our spouses and children for believing in us and giving us time and space to pursue our professional dream. Without their support and care, we would not be where we are now. Special gratitude to our siblings and extended family members who always challenged us to be better at what we do.

Once again, we like to express our sincere gratitude to our beloved parents, who inspired us to become a better human being while striving for professional and personal success.

Contents

About the Authors

Abdullah G. Alharbi received his B.Sc. degree in Electronics and Communications Engineering in 2010 from Qassim University, Saudi Arabia. He got his M.S. and Ph.D. degrees in Electrical and Computer Engineering in 2014 and 2017, respectively, from the University of Missouri - Kansas City (UMKC). Currently, he is an Assistant Professor in the Department of Electrical Engineering at Al Jouf University in Saudi Arabia. From 2010 to 2012, he worked in Saudi Aramco Company as an Electrical Engineer. He authored ~15 journal and conference papers and book chapters. He is a member of IEEE, IEEE Circuits and Systems Society, IEEE Young Professionals, IEEE Signal Processing Society, IEEE Instrumentation and Measurement Society, ACM, and Gulf Engineering Union. His research interests include digital IC design, memristor, and emerging memory devices.

Masud H. Chowdhury received his B.Sc. degree in Electrical and Electronic Engineering from Bangladesh University of Engineering and Technology (BUET), Dhaka, in 1998 and his Ph.D. degree in Computer Engineering from Northwestern University, Evanston, Illinois, USA, in 2004. Currently, he is the Associate Dean of the School of Computing and Engineering and a Professor in the Department of Computer Science Electrical Engineering at UMKC. He has published more than 170 articles in various journals and conferences in the fields of microelectronics and nanotechnology. Dr. Chowdhury has served as Chair of the IEEE VLSI Systems and Applications Technical Committee from 2014 to 2016. He has been serving in the Editorial Boards of *IEEE TCAS II*, *IEEE TVLSI*, *IEEE JETCAS*, *Springer Journal of Circuit, Systems, and Signal Processing*, and *Elsevier Microelectronics Journal*. He has also been serving the professional community as symposium chair, conference track chair, special session organizer, review committee chair, session chair, and many other roles for the last 15 years. He has supervised more than 60 Ph.D. and M.S. students and a couple of postdoctoral fellows. Dr. Chowdhury received the *Leadership Excellence Achievement Program (LEAP) Award 2017* from the Missouri Society of Professional Engineers (MSPE) for demonstrating mentoring abilities that encourage students to seek leadership excellence in the engineering profession.

List of Figures

List of Tables

List of Abbreviations and Symbols

R	Resistor
C	Capacitor
L	Inductor
M	Memristor
D	Diode
R_M	Memristance
G_M	Memductance
CMOS	Complementary Metal-Oxide-Semiconductor
VM	Volatile Memory
NVM	Non-Volatile Memory
SRAM	Static Random Access Memory
DRAM	Dynamic Random Access Memory
ReRAM	Resistive Random Access Memory
MTJ	Magnetic Tunneling Junction
MRAM	Magnetic Random Access Memory
PCM	Phase Change Memory
I	Electrical Current
V	Electrical Voltage
AC	Alternating Current
DC	Direct Current
q	Electrical Charges
HP	Hewlett-Packard.
TiO_2	Titanium Dioxide
MoS_2	Molybdenum Disulfide
STBM	Simmons Tunneling Barrier Model
TEAM	ThrEshold Adaptive Memristor Model
RMSE	Root-Mean-Square Error
GMM	Generalized Memristor Model
VTEAM	Voltage ThrEshold Adaptive Memristor
CCII	Second-Generation Current Conveyor
CFOA	Current Feedback Operational Amplifier

Op-Amps	Operational Amplifiers
LDR	Light Dependent Resistor
WO	Wien Oscillators
ϕ	Magnetic Flux
$p(t)$	Instantaneous Power
μ_v	Mobility Factor
R_{on}	Minimum Memristance
R_{off}	Maximum Memristance
$f(x)$	Window Function
V_{in}	Input Voltage
V_o	Output Voltage
V_{fb}	Feedback Voltage
V_T	Thermal Voltage
I_{ES}	Saturation Current
ψ	Shaping Function
$\psi(i)$	Current Shaping Function
α	Multiplier Constant
i_{off}	Threshold Current
i_{on}	Threshold Current

Chapter 1
Memristor Theory and Concepts

1.1 Introduction

An electric circuit is a closed loop that consists of passive and/or active electrical elements [1]. There are three fundamental passive elements – the resistor (R), the capacitor (C), and the inductor (L), whose properties, impacts, and applications in the electrical circuits and systems are well defined and understood. These two-terminal passive elements are represented in terms of three relationships between two of the four basic circuit variables (electrical quantities). The four fundamental circuit variables are known as the voltage (v), the current (i), the electric charge (q), and the magnetic flux (ϕ), as shown in Fig. 1.1b. From the symmetry arguments in Fig. 1.1b, there are six combinations of the four basic circuit variables, and five of them are well understood. In other words, three of these combinations define the three fundamental passive elements (the resistor, the capacitor, and the inductor), and two of these combinations are defined by Faraday's law of induction (the charge is the time integral of the current, and the flux is the time integral of the voltage). One of these combinations was undefined before 1971. It is the relationship between the electric charge (q) and the magnetic flux (ϕ). Professor Leon Chua found that the missing link between the charge (q) and the magnetic flux (ϕ) could be linked by a new and the fourth passive element, namely, the memristor (M), which was presented in his seminal paper [2] in 1971.

After arranging the four fundamental circuit variables in symmetry arguments, Leon Chua compared this arrangement to Aristotle's theory of matter, as shown in Fig. 1.1a. He was consequently able to postulate the existence of the new and the fourth passive element from this arrangement. More details about his innovation can be found in his lecture delivered at the *Memristor and Memristive Systems Symposium at the University of California Berkeley on November 21, 2008* [4, 5].

It is important to mention that Widrow and Hoff proposed a three-terminal element named "memistor" in 1960 to be used in neural networks [6]. The problem with this element is that its behavior cannot be predicted and controlled when it is

© Springer Nature Switzerland AG 2021
A. G. Alharbi, M. H. Chowdhury, *Memristor Emulator Circuits*,
https://doi.org/10.1007/978-3-030-51882-0_1

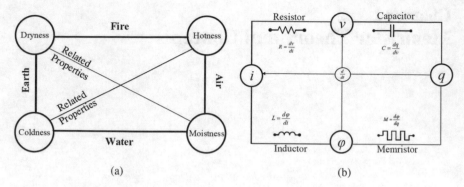

Fig. 1.1 (**a**) Aristotle's theory of matter (adapted from [4, 5]) and (**b**) four fundamental passive elements. (Adapted from [2, 3])

connected to an electrical circuit. Also, "memristor" and "memistor" are different circuit elements. In other words, the memristor is a two-terminal element, and the memistor is a three-terminal element. More details about why memristor and memistor are different can be found in [7]. The memristor and the behavior of the three fundamental passive elements in the I-V plane are defined in the next subsection.

1.2 What Is the Memristor?

A memristor (or memory resistor) is a two-terminal passive element with a unique nonlinear feature, which is not observed in other two-terminal elements like resistors, inductors, or capacitors. Even though the behavior of memristors was investigated two centuries ago [8], the idea of implementing a memristor was theoretically proposed by Professor Leon Chua in 1971 for the first time [2]. He showed that no combination of the well-known two-terminal passive elements (R, L, and C) could duplicate the characteristics of a memristor (M). Also, the memristor is the only two-terminal element that can show the missing link between the flux and the electrical charge.

$$M = \frac{d\phi}{dq} \tag{1.1}$$

In the I-V plane, the memristor shows a unique pinched hysteresis loop with frequency dependence, as shown in Fig. 1.2. It is a special type of resistor, where the resistance (R_{on} and R_{off}) increases/decreases depending on the polarity of either the current passing through it or the applied voltage. The removal/zero value of the excitation source does not change the memristor resistance (memristance). This

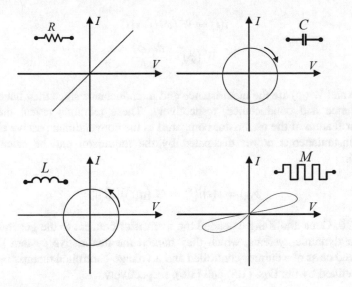

Fig. 1.2 The current-voltage characteristic of the four fundamental passive elements (R, C, L, and M) with a sinusoidal input voltage. (Adapted from [9])

property gives the memristor the capability to act as a memory. Professor Chua defined the memristor (M), with memristance as in Eq. (1.1) [2].

The behaviors of the four fundamental passive elements (R, C, L, and M) in the I-V plane with a sinusoidal input voltage are shown in Fig. 1.2. It is well-known that the resistor shows a linear relationship between the voltage and the current, while the energy-storing elements (C and L) exhibit a circular shape with ±90 phase difference [9]. However, the memristor shows a unique pinched hysteresis loop with zero-crossing in the I-V plane. Therefore, from Fig. 1.2, it can be concluded that the memristor could be the fourth fundamental passive element since it exhibits a unique behavior in the I-V plane that has not been duplicated or seen before in the three well-known two-terminal passive elements (R, L, and C).

The memristor can be of two types: (i) charge-dependent or current-controlled and (ii) flux-dependent or voltage-controlled. If the memristance relation is a single-valued function of the charge, it is called a charge-controlled memristor. Similarly, if the memristance relation is a single-valued function of the flux, it is called a flux-controlled memristor [2, 5, 10]. The voltage across a charge-controlled memristor is represented by Eq. (1.2).

$$v(t) = M\left(q(t)\right)i(t) \tag{1.2a}$$

$$M\left(q\right) = \frac{d\phi(q)}{dq} \tag{1.2b}$$

The current of a flux-controlled memristor is represented by Eqs. (1.3).

$$i(t) = W(\phi(t)) v(t) \tag{1.3a}$$

$$W(\phi) = \frac{dq(\phi)}{d\phi} \tag{1.3b}$$

$M(q)$ and $W(\phi)$ are the memristance and memductance since they have the unit of resistance and conductance, respectively. These relations reveal the unique nonlinear feature of the memristor compared to the conventional passive elements.

The instantaneous power dissipated by the memristor can be calculated by Eq. (1.4).

$$p(t) = v(t)i(t) = M(q(t))[i(t)]^2 \tag{1.4}$$

In 1976, Chua and Kang extended the memristor concept to the general class of nonlinear dynamic systems, which they named the memristive system [11]. The generalized class of a current-controlled and a voltage-controlled memristive system can be defined by the Eqs. (1.5) and (1.6), respectively.

$$v(t) = R(x, i(t), t)i(t) \tag{1.5a}$$

$$\frac{dx}{dt} = f(x, i(t), t) \tag{1.5b}$$

$$i(t) = G(x, v(t), t)v(t) \tag{1.6a}$$

$$\frac{dx}{dt} = f(x, v(t), t) \tag{1.6b}$$

where $v(t)$ and $i(t)$ are the input voltage and the input current. Here, the variable x is the state variable, and the functions f, R, and G are explicit functions of time. Besides, it is important to mention that the above equations opened the door to model most of the nonvolatile memory devices or the memristive devices in recent times [10, 12].

1.3 Memristor Fingerprints

For a device to be considered as a memristor, it must have three significant fingerprints [5, 10, 13, 14]. Therefore, any memristor emulator circuit must also comply with these three fingerprints or defining characteristics. In this section, we briefly summarize these fingerprints to establish the intellectual merits of the proposed emulator circuits in Chaps. 3, 4, 5, and 6. These three fingerprints are as follows:

1. Memristor Fingerprint 1 – *pinched hysteresis loop*: The first significant signature of the memristor is the unique pinched hysteresis loop that distinguishes it from any device that is not memristive in the I-V plane.

   (a) In the I-V plane, the Lissajous figure of all memristors, having positive memristance and operated by a sinusoidal signal of any amplitude and frequency, must go through the origin.
   (b) The value of $v(t)$ and $i(t)$ in the Lissajous figure should be the same only when it will pass through the origin. However, for the rest of the times, the voltage (v) and the current (i) should have different values.

2. Memristor Fingerprint 2 – *decrease of the hysteresis loop area with the increase of the frequency*: The second vital signature of the memristor is the inversely proportional relationship between the frequency of the periodic operating signal and the memristor's hysteresis lobe area. It states that with the increment of frequency, the lobe area will decrease.
3. Memristor Fingerprint 3 – *no loop at the infinite frequency*: With the increase of frequency, at some point, the lobe area will reduce. At higher frequencies, a memristor gradually loses its unique nonlinearity, and there wouldn't be any loop above a certain frequency. The values of the voltage (v) and the current (i) remain on the same line for all times in the I-V plane above that frequency, which means that the memristor behaves as a linear resistor above a certain frequency.

1.4 HP Memristor

In 2008, S. Williams and his team from the HP laboratory were the first group to fabricate the material structure of a memristor using titanium dioxide (TiO_2). They introduced the simplest model for the memristor [3, 15] (see Fig. 1.3). Before the introduction of this device, the memristor was only a theoretical concept. The HP

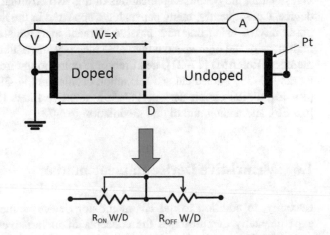

Fig. 1.3 Memristive device proposed by the HP Laboratory and its equivalent circuit. (Adapted from [3])

memristor was the first two-terminal nanoscale device that showed a pinched hysteresis loop in the I-V plane, where the relation between the current and the voltage can be given by Eq. (1.7).

$$v(t) = \left(R_{off} - \left(R_{off} - R_{on}\right) x\right) i(t) \tag{1.7a}$$

$$\frac{dx}{dt} = k\, i(t) f(x) \tag{1.7b}$$

Here, $i(t)$ is the current passing through the memristor, $v(t)$ is the voltage across the memristor, R_{on} and R_{off} are the minimum and the maximum obtainable memristance, x is the state variable of the memristor that is limited by $(0, 1)$, D is the total length of the TiO_2 layer, $k = \mu_v(R_{on}/D)$, μ_v is the mobility factor, and $f(x)$ is the window function. By differentiating Eq. (1.7a) and substituting it into Eq. (1.7b), we can find the rate of change in the memristance (Rm) as in Eq. (1.8). Here, Rd is the difference between the minimum and the maximum obtainable memristance.

$$\frac{dR_m}{dt} = -kRd i(t) f(x) \tag{1.8}$$

It was observed that the rate of change of the memristance (Rm) is linearly proportional to the current passing through the memristor. Before ending this section, it is worth mentioning that the switching phenomena in the TiO_2 thin film were reported before HP's memristive device in [16].

1.5   Memristor Applications

After the introduction of the HP memristor model, many engineers and scientists from all over the world have shown considerable interest in understanding and investigating the potential applications of this two-terminal nonlinear and nanoscale device. Consequently, many papers have appeared in the literature that explores the possibility of using this new passive element in various applications that include high-density and high-speed nonvolatile memory arrays like resistive random access memory (ReRAM) [17–21], neuromorphic computing circuits [22–28], neural networks [29–33], sinusoidal and relaxation oscillators [34–39], analog circuit designs [40–46], digital circuit designs [47–56], adaptive filters [57–63], chaotic systems [62–66], and analog and digital modulation [67–73].

1.6   Memristive Device Implementation

Recently, in addition to the HP memristor device, some other researchers have experimentally demonstrated the concepts of memristive devices using different materials. Some examples of the attempted memristive device implementations are ferroelectric memristor [74], resistive switching in silicon suboxide films [75, 76], polymer memristor [77, 78], resonant-tunneling diode memristor [79], spintronic memristor [80], gate-tunable memristive phenomena mediated by grain boundaries in a single-layer molybdenum disulfide (MoS_2) [81], and memristive devices based on graphene oxide [82] and carbon nanowalls [83]. More details can be found in [84–86]. However, all these memristive devices are not physically available in the commercial design houses due to the complexity and high cost of fabricating them in nanoscale dimensions. Besides, the research community is still trying to understand different physical and electrical characteristics of the memristors and figure out what would be the feasible materials to achieve these properties in a reliable device. Therefore, most of the research on memristor is still purely at the theoretical and analytical (or rudimentary testing) stages in the academic laboratories.

1.7   Different Models of Memristor and the Importance of Memristor Emulator Circuits

Considering the enormous potentials of the memristors, there has been a surge of interest in performing different theoretical and experimental works on memristors. However, since there is no practical memristive device available to integrate with any on-chip application, numerous circuit-based emulators and macro-models are being developed to investigate and understand the properties and the potential applications of the memristors. These circuit-based models or the emulator circuits can mimic some of the perceived behaviors of the memristor. Most of the available emulator circuits and macro models are based on the HP memristor, which is a linear model. Although the HP model enabled researchers to understand and analyze some aspects of the memristor, it was later acknowledged that the HP model failed to provide the actual nonlinear behaviors of the memristor. Besides, the cost and the complexity of fabricating a TiO_2-based nanoscale memristor prevented both the industry and the academic communities from generating real test data for the validation of the perceived properties and applications. Therefore, as an improvement over the HP model, more precise memristor models (mathematical and analytical models) were recently proposed for both the voltage-controlled and the current-controlled memristors. These are the Simmons Tunneling Barrier Model (STBM) [87], the ThrEshold Adaptive Memristor (TEAM) Model [88], the Generalized Memristor Model (GMM) [89, 90], and the Voltage ThrEshold Adaptive Memristor (VTEAM) [91]. However, to the best of our knowledge, there is no efficient emulator circuit to mimic the electrical nonlinear behavior of the memristors defined by the STBM, TEAM, GMM, and VTEAM models.

1.8   Conclusion

In this book, the concepts and the designs of a set of generic and practical memristor emulator circuits have been presented. The proposed emulator circuits can be used to analyze and understand the nonlinear behavior of the well-known current-controlled and voltage-controlled memristor models. Besides, the proposed emulator circuit development techniques can be configured for both the floating and the grounded memristor models. It is important to note that memristor is perceived as a two-terminal device. Therefore, it may appear in a circuit both in the floating or grounded states. The proposed memristor emulator circuits are more straightforward and can be utilized for understanding the nonlinear behaviors of the memristors for different application environments. Therefore, these emulators can overcome the limitation of the existing emulators that cannot mimic the nonlinear behavior. In addition to the mathematical modeling and analysis of the proposed emulators, SPICE simulations and experimental results are provided. The efficacy of the proposed emulators has been verified by using the memristor circuit model in designing some applications like Wien Bridge oscillators. A brief comparison with the previously published emulators is included to highlight the advantages of the proposed emulator designs.

Chapter 2
Memristor Models and Emulators: A Literature Review

2.1 Memristor Models

2.1.1 Linear Ion Drift Model

It is observed that the rate of change of the memristance (R_m) in HP's device is linearly proportional to the current passing through the memristor (see Sect. 1.4 of Chap. 1). Researchers found that this model has some limitations that need to be resolved. For instance, the state variable of the device has a boundary effect, and the large value of the electric field inside this device is not considered. Since a practical memristor is a nonlinear device, the original linear HP model will not be suitable for developing a practical emulator circuit or application. To overcome some of the drawbacks of this device, researchers suggested using a nonlinear window function $f(x)$ to add nonlinear effects to the HP model. New nonlinear memristor models were also proposed in literature based on the Eqs. (1.5 and 1.6), which are discussed in Sect. 1.2 of Chap. 1.

2.1.2 Nonlinear Window Function

Joglekar and Wolf proposed a window function with a controlled parameter, p [92], to be added to the HP model to include nonlinearity in the behavior. When p increases in Eq. (2.1), the model becomes linear. Here, p is a positive integer, and x is a state variable of the memristor.

$$f(x) = 1 - (2x - 1)^{2p} \tag{2.1}$$

© Springer Nature Switzerland AG 2021
A. G. Alharbi, M. H. Chowdhury, *Memristor Emulator Circuits*,
https://doi.org/10.1007/978-3-030-51882-0_2

Practically, the window function in Eq. (2.1) had some limitations and could not solve the boundary effect. This motivated Biolek et al. in [93] to propose another window function as in Eq. (2.2).

$$f(x) = 1 - (x - stp(-i))^{2p} \tag{2.2a}$$

$$stp(-i) = \begin{cases} 1, & i \geq 0 \\ 0, & i < 0 \end{cases} \tag{2.2b}$$

where i is the memristive device current and stp is the step function. This window function is more suitable for the memristor device since it depends on both the state variable (x) and the current (i). However, these earlier window functions do not provide a scale factor or threshold technique and, therefore, cannot be controlled [88]. To overcome these drawbacks, an alternative window function with the scale factor was introduced by Prodromakis et al. in [94], as illustrated in Eq. (2.3).

$$f(x) = j\left(1 - \left[(x - 0.5)^2 + 0.75\right]^p\right) \tag{2.3}$$

where j is a scalar that acts as the second control parameter, which controls the maximum value of the window function, $f(x)$. In this window function, if $p = 1$, it matches the HP model. However, the current (i) is not accounted for in Eq. (2.3). This window function is also not suitable for programming an analog circuit due to the issue of a boundary lock. Zha et al. [46] presented a new window function to overcome the limitations of the previous window functions. Both the current and the controlled parameters are considered in the window function of [46], as shown in Eq. (2.4).

$$f(x) = 1 - [0.25\,(x - stp(i))^2 + 0.75]^p \tag{2.4}$$

A comparison among the above window functions can be found in detail in [46, 88, 95].

2.1.3 Nonlinear Ion Drift Model

Yang et al. proposed an exponential model based on the experimental results in [96]. This model has been widely used in memristor applications in recent times. In this model, the relationship between the current and the voltage is as shown in Eq. (2.5).

$$i(t) = x(t)^n \beta \sinh\left(\alpha v(t)\right) + \chi\left[e^{\gamma v(t)} - 1\right] \tag{2.5a}$$

$$\frac{dx}{dt} = a.v(t)^m f(x) \tag{2.5b}$$

where $i(t)$ and $v(t)$ are the input current and voltage, respectively. β, χ, and γ are the experimental fitting parameters, and n is a controlled parameter that controls the state variable (x) of the memristor. Here a and m are constants, and $f(x)$ is a window function.

2.1.4 Simmons Tunnel Barrier Model (STBM)

This model considers that the switching behavior is nonlinear asymmetric due to an exponential dependence of the movement of the ionized dopants that corresponds to the state variable. In [87], the STBM describes the switching behavior of the memristors, where the state variable (x) is described as in Eq. (2.6), where C_{off}, C_{on}, i_{off}, i_{on}, a_{off}, a_{on}, b, and w_c are the fitting parameters and exp is the exponential function. Kvatinsky et al. [88] simplified this equation to separate the variables under certain conditions. Therefore, the variable is approximated to $\frac{dx}{dt} = g(i)f(x)$, where g (i) is shown in Eq. (2.7) and $f(x)$ is the window function of the memristor.

$$\frac{dx}{dt} = \begin{cases} C_{off} \sinh\left(\dfrac{i}{i_{off}}\right) \exp\left[-\exp\left(\dfrac{x - a_{off}}{w_c} - \dfrac{|i|}{b}\right) - \dfrac{x}{w_c}\right], & i > 0 \\[3mm] C_{on} \sinh\left(\dfrac{i}{i_{on}}\right) \exp\left[-\exp\left(\dfrac{x - a_{on}}{w_c} - \dfrac{|i|}{b}\right) - \dfrac{x}{w_c}\right], & i < 0 \end{cases} \tag{2.6}$$

$$g(i) = \begin{cases} C_{off} \sinh\left(\dfrac{i}{i_{off}}\right), & i > 0 \\[3mm] C_{on} \sinh\left(\dfrac{i}{i_{on}}\right), & i < 0 \end{cases} \tag{2.7}$$

2.1.5 ThrEshold Adaptive Memristor Model (TEAM)

Kvatinsky et al. [88] proposed the TEAM model, which is a simplified memristor model to fit different fabricated models and to support simple analysis and computational efficiency. According to their model, the rate of change of the state variable is shown in Eq. (2.8).

$$\frac{dx}{dt} = \begin{cases} k_{off}\left(\dfrac{i(t)}{i_{off}} - 1\right)^{\alpha_{off}} f_{off}(x), & i_{off} < i \\ 0, & i_{on} < i < i_{off} \\ k_{on}\left(\dfrac{i(t)}{i_{on}} - 1\right)^{\alpha_{on}} f_{on}(x), & i < i_{off} \end{cases} \qquad (2.8)$$

Here, f_{off} and f_{on} are off and on switching window functions, respectively, that describe the physical nonlinearity of the device. k_{off}, k_{on}, α_{off}, and α_{off} are constants, and i_{off} and i_{on} are the off and on threshold currents, respectively.

2.1.6 Generalized Memristor Model (GMM)

The generalized model was built based on a hyperbolic sinusoid, which is modulated with the state variable, $I(t)$ [89, 90]. This hyperbolic sinusoid shape is chosen to model the metal-insulator-metal (MIM) structure of the memristors [90]. The parameters a_1, a_2, and b are selected to fit different device structures of the memristors. The following general Eq. (2.9) shows the change in the state variable as in Eqs. (2.10) and (2.11).

$$I(t) = \begin{cases} a_1 x(t) \ \sinh(bV(t)), & V(t) \geq 0 \\ a_2 x(t) \ \sinh(bV(t)), & V(t) < 0 \end{cases} \qquad (2.9)$$

$$\frac{dx}{dt} = \eta g(V(t))f(x(t)) \qquad (2.10)$$

$$g(V(t)) = \begin{cases} A_p\left(e^{V(t)} - e^{Vp}\right), & V(t) > Vp \\ -A_n\left(e^{-V(t)} - e^{Vn}\right), & V(t) < -Vn \\ 0, & -Vn \leq V(t) \leq Vp \end{cases} \qquad (2.11)$$

Here, $f(x(t))$ is a sophisticated window function, V_p and V_n represent the positive and negative threshold voltages of the memristor, and A_p and A_n are the magnitudes of the exponential that can be adjusted to control the speed of changing the state after exceeding the threshold.

2.1.7 A General Voltage-Controlled VTEAM Model

In the Voltage ThrEshold Adaptive Memristor (VTEAM) model [91], the rate of change in the state variable is shown in Eq. (2.12), where k_{off}, k_{on}, α_{off}, and α_{on} are the constants and are chosen to fit the experimental results. v_{off} and v_{on} are two threshold voltages, and $f_{on}(x)$ and $f_{off}(x)$ represent the window functions.

$$\frac{dx}{dt} = \begin{cases} k_{off}\left(\frac{v(t)}{v_{off}} - 1\right)^{\alpha_{off}} f_{off}(x), & 0 < v_{off} < v \\ 0, & v_{on} < v < v_{off} \\ k_{on}\left(\frac{v(t)}{v_{on}} - 1\right)^{\alpha_{on}} f_{on}(x), & v < v_{on} < 0 \end{cases} \qquad (2.12)$$

As opposed to the HP model, the models illustrated in this section can represent the physical nonlinearity of a memristor. However, although these models accurately describe the theoretical aspects of a solid-state memristor device, we still must rely on emulator circuits or macro-models due to the lack of any real memristor.

2.2 Memristor SPICE Models

To validate the applications of a memristor, a more precise and realistic memristor model is needed to implement a nonlinear memristor and investigate all the anticipated applications. Hence, to study the complex and dynamic nature of this device, numerous circuit-based emulators and macro-models are proposed in the literature. For example, a SPICE modeling approach for memristors based on the Simmons Tunneling Barrier Model (STBM) was presented in [97]. However, this model is more complicated, and as a result, a modified version was introduced in [98] to increase the performance of the model in the SPICE environment. Biolek et al. also proposed a SPICE model of memristor with nonlinear dopant drift [93], and this model is more accurate and widely used in the literature. Another SPICE modeling of nonlinear memristive behavior was presented in [99, 100]. These models are very useful in digital memristor applications since they have the threshold technique. A generalized memristive device SPICE model was proposed in [101]. An approximated SPICE model for memristor was presented in [102]. Recently, a SPICE implementation of a novel compact model for nonlinear memristive devices, which used the memdiode model, was introduced in [103]. A SPICE compact modeling of bipolar/unipolar memristor switching was proposed in [104]. However, some of these models have many limitations and cannot mimic the physically developed memristor. Most of these models are only applicable to a computer-aided simulation of a memristor. Therefore, there is a critical need to develop more realistic circuit-based emulators to analyze and understand the properties of the nonlinear memristor. In the next section, a brief review of some important emulator circuits is presented.

2.3   Memristor Emulator Circuits

With the surge of interest to understand the nonlinear properties and investigate potential applications of memristor, many initiatives to develop memristor emulators have been taken. Due to the complexity and high cost of fabricating nanoscale memristor prototypes and solid-state memristive devices, the research community focuses on developing emulator circuits to mimic the dynamic behavior of the memristors. For low-cost samples and educational purposes, most of the research works in this area are still based on memristor emulator circuits that can be either current-controlled or voltage-controlled. Therefore, many emulator circuits have been proposed based on different design methodologies and using off-the-shelf active and passive circuit components that are commercially available. The existing memristor emulators can be divided into two groups: analog emulators and digital emulators [105]. The first analog emulator that appeared in the literature was introduced by Professor Chua in 1971 [2].

Some of the emulators are implemented using two methods based on the general model of the memristor in Fig. 2.1 [105]. The first method, shown in Fig. 2.1a, works by integrating the input signal such as voltage or current. For determining whether the memristance is incremental or decremental, we must apply the nonlinear function block $f(x)$. Then, we must differentiate the output of the nonlinear function back to the voltage or current. The second method, which is shown in Fig. 2.1b, also works by integrating the input signal, such as the voltage or the current. Then, the input signal transforms through a nonlinear function $g(x)$. The output of the nonlinear function and the input signal are multiplied together. The output of this multiplication represents the current or voltage of the memristor. More details about these methods can be found in [105].

Several emulator designs, including the original one presented in [2], are available in the literature. These emulators use a large number of active and passive components such as Metal Oxide Semiconductor Field Effect Transistor (MOSFET), operational amplifier (Op-Amp), second-generation current conveyor (CCII), transistor, analog multiplier, capacitor, inductor, resistor, potentiometer, Junction Field-

Fig. 2.1 General models of analog emulators. (Adapted from [105])

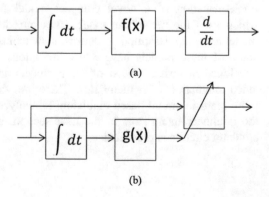

(a)

(b)

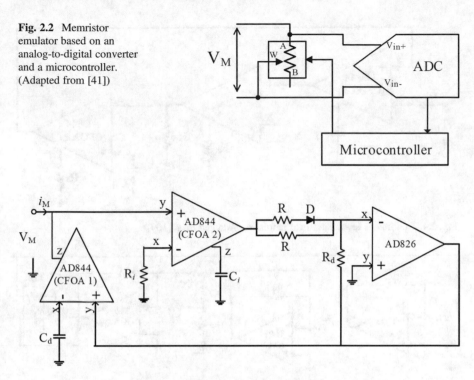

Fig. 2.2 Memristor emulator based on an analog-to-digital converter and a microcontroller. (Adapted from [41])

Fig. 2.3 Memristor emulator based on nonlinear resistor. (Adapted from [108])

Effect Transistor (JFET), Zener diodes, bipolar junction transistor (BJT), diodes, differential difference current conveyors (DDCC), current feedback operational amplifier (CFOA), microcontroller unit, analog-to-digital converter (ADC), and digital-to-analog converter (DAC). For instance, the memristor emulator introduced in [41] used a microcontroller unit, an ADC, a DAC, and a potentiometer, as shown in Fig. 2.2. Another hybrid emulator circuit, which used an ADC and a DAC, is introduced in [106, 107]. These emulator circuits are topologically complex, which limits their applications because it is difficult to connect those with other active and passive devices.

$$R_{in} = \frac{R_i C_i}{R_d C_d} \frac{d_{vR}}{d_{iR}} R_{dif} \qquad (2.13)$$

The memristor emulator introduced in [108] used two CFOAs, one voltage-feedback operational amplifier, some passive elements, and a light dependent resistor (LDR). In this emulator, the nonlinear property of memristor was obtained using one LDR (LA-541B) and two resistors, as shown in Fig. 2.3. The CFOA-1 operates as the integrator, while the CFOA-2 operates as the differentiator. The memristance of this memristor emulator is represented by Eq. (2.13). The emulator circuit presented in [109] used two CFOAs, three resistors, two grounded capacitors, and

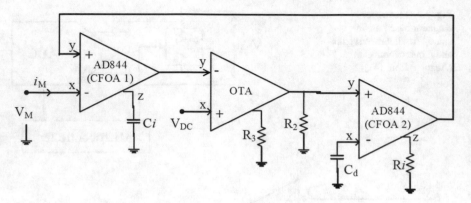

Fig. 2.4 Continuous level memristor emulator. (Adapted from [109])

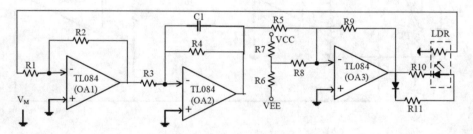

Fig. 2.5 Memristor emulator based on LDR. (Adapted from [117])

one operational transconductance amplifier (OTA), as shown in Fig. 2.4. According to the authors, this emulator is a continuous level emulator that can have more than two state values. The voltage, V_{DC} is used to make the OTA operate in the nonlinear region. This emulator is suitable for the analog application since it has a continuous level state, which is very useful in oscillator applications. The memristance (M) of this emulator is, as shown in Eq. (2.14).

$$M = \frac{C_d R_i R_3}{R_i R_{eq}} \tag{2.14}$$

In [110], another memristor emulator is proposed, which is composed of an adder, ten transistors, five OP-Amps, and eight resistors. An electromechanical emulator of memristive systems was introduced in [111]. Cubic flux-controlled emulators were introduced in [112–114] based on the cubic nonlinearity illustrated in [115]. The emulator circuit presented in [67] used two CFOAs, one diode, four resistors, and two grounded capacitors. Some voltage and current-controlled memristor emulators were presented in [116]. The memristor emulator introduced in [117] used three Op-Amps, one diode, a large number of passive elements, and a light-dependent resistor (LDR), as in Fig. 2.5.

Fig. 2.6 A floating
memristor emulator.
(Adapted from [120])

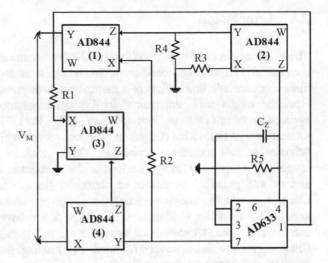

A memristor emulator circuit based on an OTA was proposed in [118]. However, the experimental results of this emulator circuit do not satisfy the condition of the memristor. A memristor emulator circuit based on an exponential amplifier and a CCII was presented in [119]. The memristor emulator presented in [120] used five CCII+, an analog multiplier, five resistors, and one capacitor, as shown in Fig. 2.6. This emulator represents a floating memristor emulator for a current-controlled memristor, where the memristance can be given by Eq. (2.15).

$$M = R_1 + \frac{R_1 R_i R_4}{R_2 R_3 C_Z} \ \phi(t) \tag{2.15}$$

The emulator presented in [121] used one CCII, one current follower, one inductor, and a nonlinear resistor. The emulator circuit proposed in [122] required four resistors, one differential difference current conveyors (DDCC), one grounded capacitor, and one analog multiplier. A CMOS-based memristor emulator was also introduced in [123–125]. An emulator based on two CCIIs, two diode-connected transistors, and one resistor is presented in [126], and the improvement of this design is shown in [127]. A floating memristor, which is an emulator-based relaxation oscillator, was presented in [128], and in [129], an approach to reduced and improved the size of this emulator was shown. A simple floating and grounded voltage-controlled emulator was presented in [129]. There are other floating emulator circuit designs available in the literature [130–134]. A phase-change memory cell-based emulator circuit was introduced in [135].

2.4 Conclusion

The emulator circuits highlighted in Sect. 2.3 have some drawbacks. Some are very complex and require rigid conditions. Some emulators do not exhibit or satisfy the three characteristic fingerprints of a memristor, as discussed in [13]. Some of these emulator circuits can be customized for different memristor models by selecting the appropriate circuit elements introduced in [127, 136, 137]. In this book, we aim to present several innovative designs of practical emulator circuits that can mimic the behavior of real memristor properties. These emulator circuits can be used to investigate many potential applications of the memristor. In Chap. 3, we propose a generic and practical memristor emulator for the current-controlled models. In Chap. 4, a simple memristor emulator design based on a single CCII and an exponential amplifier is illustrated. In Chap. 5, we have shown the design of a generic and practical memristor emulator for the voltage-controlled models. In Chap. 6, we present a novel grounded and floating flux-controlled memristive emulator suitable for analog applications.

Chapter 3
Generic and Practical Emulators for the Current-Controlled Memristor Models

3.1 Proposed Practical Model and Memristor Emulator Circuit

Considering the limitation of the original linear HP model, we intend to develop a practical model and an emulator circuit implementation technique that can be applied to the nonlinear models (STBM and TEAM). For simplicity, we assumed that the window function, $f(x)$, in these models is linear and equal to 1. This assumption is based on the mathematical analysis presented in [87, 88]. With this assumption, the rate of change of memristance is proportional to a function of the current passing through the memristor. Therefore, the memristance can be represented by Eq. (3.1), where $\psi(i)$ is the current shaping function and i_{off} and i_{on} are the threshold currents and can be made equal to zero to fit the STBM.

$$\frac{dR_m}{dt} = \psi(i) = \begin{cases} \psi_{off}(i), & i_{off} < i \\ 0, & i_{on} < i < i_{off} \\ \psi_{on}(i), & i < i_{on} \end{cases} \tag{3.1}$$

Our goal is to build an emulator for the memristor, where the memristance changes according to Eq. (3.1), and the current-voltage relation follows Eq. (1.5) from Chap. 1. Figure 3.1 shows the proposed emulator that consists of two second-generation current conveyors (CCII) and a multiplier, in addition to a block that represents the current shaping function. The current conveyors are labeled U1 and U2. The input voltage to the circuit is given by Eq. (3.2), where V_{fb} is the feedback voltage, which is the voltage multiplication of V_Z and the time integral of V_S. Hence, V_{fb} can be expressed in terms of V_Z and V_S as in Eq. (3.3). Here α is the multiplier constant, and ψ is the shaping function. The output voltage of the first CCII (U1) is $V_Z = iinR1$. Therefore, the feedback voltage can be reduced to Eq. (3.4).

© Springer Nature Switzerland AG 2021
A. G. Alharbi, M. H. Chowdhury, *Memristor Emulator Circuits*,
https://doi.org/10.1007/978-3-030-51882-0_3

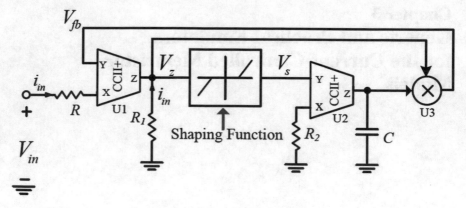

Fig. 3.1 Proposed design of the emulator circuit for the current-controlled memristor

$$V_{in} = i_{in}R + V_{fb} \tag{3.2}$$

$$V_{fb} = \frac{\alpha}{R_2 C} V_z \int_0^t V_S(\tau)\, d\tau, \; V_S = \psi(V_z) \tag{3.3}$$

$$V_{fb} = \frac{\alpha R1}{R_2 C} i_{in} \int_0^t \psi(i_{in}(\tau)R_1)\, d\tau \tag{3.4}$$

By substituting Eq. (3.4) into Eq. (3.2), the input voltage can be given by Eq. (3.5), and the input resistance can be given by Eq. (3.6).

$$V_{in} = i_{in}R + \frac{\alpha R_1}{R_2 C} i_{in} \int_0^t \psi(i_{in}(\tau)R_1)\, d\tau \tag{3.5}$$

$$R_m = R + \frac{\alpha R1}{R_2 C} \int_0^t \psi(i_{in}(\tau)R_1)\, d\tau \tag{3.6}$$

By differentiating the Eq. (3.6), we obtain the rate of change in memristance as in Eq. (3.7). This equation is similar to the required Eq. (3.1) to realize the memristor model.

$$\frac{dR_m}{dt} = \frac{\alpha R_1}{R_2 C} \psi(i_{in}(t)R_1) \tag{3.7}$$

The key component of the proposed emulator circuit of Fig. 3.1 is the shaping function. In the HP model, the shaping function is absent, and a short circuit would be the simplest shaping function (which means $\psi = V_z$) for this case. Therefore, according to the ideal HP memristor model, the rate of change in the memristance can be given by Eq. (3.8).

$$\frac{dR_m}{dt} = \frac{\alpha R_1{}^2}{R_2 C} \, i_{in} = K i_{in} \tag{3.8}$$

The realization and analysis of this model Eq. (3.8) have been verified by Elwakil et al. in [116]. The results from their implemented emulator match well with the ideal HP memristor model [3]. Since this ideal model does not represent the nonlinear nature of the practical memristor, the emulator must include the shaping function. For the emulator to fit a specific function, the shaping function should fit the Eq. (3.7) of the rate of change in the memristance. In the next two sections, we present two practical emulator circuits using our technique proposed in Fig. 3.1.

3.2   Implementation and Validation of an Emulator Circuit for the STBM

To realize a practical shaping function, a nonlinear relation is required. In the Simmons Tunneling Barrier Model (STBM), the nonlinearity is exponential. To implement exponential nonlinearity, we plan to utilize the well-known exponential voltage amplifier circuit, as shown in Fig. 3.2a. The output voltage of the exponential amplifier can be given by Eq. (3.9), where I_{ES} is the reverse saturation current.

$$V_{o1} = I_{ES} R_D \left(1 - e^{\frac{V_{in}}{V_T}}\right) \approx -I_{ES} R_D e^{\frac{V_{in}}{V_T}} \tag{3.9}$$

To achieve a *sinh* function as in the Simmons model, another amplifier circuit with a reversed diode connection is used. The two amplifier circuits are connected, as shown in Fig. 3.2b. The output voltage ($Vo1$) of the first Op-Amp can be given by Eq. (3.10). The output voltage ($Vo2$) of the second Op-Amp is given by Eq. (3.11).

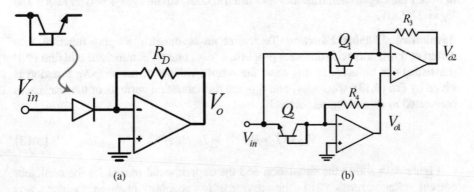

(a)   (b)

Fig. 3.2 Schematic diagrams – (**a**) basic exponential amplifier and (**b**) shaping function

$$V_{o1} \approx -I_{ES}R_4 e^{\frac{-Vin}{V_T}} \tag{3.10}$$

$$V_{o2} \approx -I_{ES}\left(R_3 e^{\frac{Vin}{V_T}} - R_4 e^{\frac{-Vin}{V_T}}\right) \tag{3.11}$$

$$V_{o2} \approx -2I_{ES}R_D \sinh\left(\frac{Vin}{V_T}\right) \tag{3.12}$$

3.2.1   Implementation of the Shaping Function

The shaping function is the critical component of our analysis and circuit development effort. The original HP linear model did not consider the shaping function. Here, we considered both symmetric and asymmetric shaping functions. To validate the proposed emulator circuit model, we have performed SPICE simulations and experimental measurements of the circuits. The circuits are implemented on a PCB board using discrete components.

Symmetric Shaping Function The characteristics of the symmetric shaping function is given by Eq. (3.12), where $R3 = R4 = RD$. The shaping function circuit of Fig. 3.2b has been implemented using an input voltage of amplitude 700 mV at 100 Hz, a DC voltage source of ± 12 V, a set of resistors of value $R3 = R4 = 10\ k\Omega$, two PN3565 transistors connected as a diode, and a TL084 operational amplifier. We used the GDS-2102 digital storage oscilloscope and Agilent 33220A waveform generator for testing and measurements. The input-output relation of the symmetric shaping circuit based on SPICE simulations and experimental measurements is shown in Fig. 3.3a. It is observed that the input-output relation is symmetric and has a dead region from around -0.6 V to $+0.6$ V, where the output is almost zero. Therefore, based on the simulation and experimental data, it can be concluded that this symmetric shaping function matches the STBM with a threshold. Moreover, to precisely model, the symmetric shaping function, a MATLAB curve fitting toolbox has been used to extract the model parameters. Figure 3.3c shows the matching between the experimental results and fitted model, where $I_{ES}R_D = 0.6912\ nV$ and $V_T = 27.7\ mV$.

Asymmetric Shaping Function To realize an asymmetric shaping function, two different transistors with different properties (e.g., one NPN transistor and one PNP transistor) can be used. In this case, the output voltage of the shaping function is given by Eq. (3.13), where I_{ES1} and I_{ES2} are the saturation currents of the transistors connected to R_3 and R_4, respectively.

$$V_{o2} \approx I_{ES2}R_4 e^{\frac{-Vin}{VT2}} - I_{ES1}R_3 e^{\frac{Vin}{VT1}} \tag{3.13}$$

Figure 3.3b shows the simulation and the experimental results for the nonlinear current-voltage relation of the asymmetric shaping function, which was implemented using two transistors (PN3565 and PN2906), an Op-Amp (TL084),

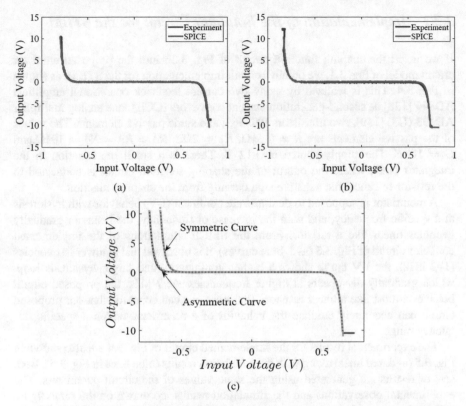

Fig. 3.3 Experimental results and SPICE simulation showing the nonlinear behavior of the proposed shaping function circuit: (**a**) symmetric shaping function, (**b**) asymmetric shaping function, and (**c**) fitted curve

and resistors of value $R3 = R4 = 1\ k\Omega$. We observed that the shape was not symmetric, and the dead region of the negative part was larger than the positive part, as shown in Fig. 3.3b. However, to have different exponents, a preamplifier can be used. We implemented the asymmetric shaping function for an input voltage of amplitude 0.7 V at 1 kHz and a DC voltage source of $\pm 12\ V$. From Fig. 3.3a, b, it can be observed that the experimental results closely matched the SPICE simulations. For the symmetrical case of Fig. 3.3a, we calculated the root mean square error (RMSE) between the experimental and simulation results. We found that the error margin was minimal and the RMSE value was 3.25e-5. To exactly model, the asymmetric shaping function, a MATLAB curve fitting toolbox was used to extract the model parameters as we did for the symmetric shaping function. Figure 3.3c shows the matching between the experimental results and fitted model of the asymmetric shaping function circuit, where $I_{ES1}R_3 = 0.6912nV$, $I_{ES2}R_4 = 0.3645nV$, $V_{T1} = 27.7mV$, and $V_{T2} = 30.3mV$. Figure 3.3c also shows the difference between the symmetric and the asymmetric shaping functions. The two shaping functions have the same right part and different left parts.

3.2.2 Implementation of the Emulator Circuit for the STBM

If we insert the shaping function circuit of Fig. 3.2b into the proposed emulator circuit model of Fig. 3.1, we obtain a circuit implementation for the STBM, as shown in Fig. 3.4. This is realized by using two current feedback operational amplifiers AD844 [138] as second-generation current conveyors (CCII), one analog multiplier AD633 (U3) [139], two transistors PN3565, and some passive elements. The values of the passive elements are $R = 2.5k\Omega$, $R1 = 2k\Omega$, $R2 = R3 = R4 = 10k\Omega$, and $C = 10nF$. The supply source is ± 12 V. There is a small modification in the integrator (U2), where the output of the shaping function circuit is connected to the resistor to cancel the negative sign coming from the shaping function.

A memristor is supposed to demonstrate nonlinear (I-V) behavior with hysteresis at a specific frequency, and with the increase of frequency, the behavior gradually becomes linear like a resistor. From the SPICE simulations of the implemented emulator circuit of Fig. 3.5 (a–d, blue curves), it is observed that at lower frequencies (1–2 kHz), the I-V curve shows a higher nonlinearity and larger hysteresis loop, which gradually disappears at higher frequencies. At 7 kHz, the proposed circuit behaves almost like a linear resistor. Therefore, we can conclude that our proposed circuit can accurately emulate the behavior of a memristor within a specific frequency range.

The experimental results for the implemented circuit of Fig. 3.4 are also shown in Fig. 3.5 (a–d, red lines) along with the simulation results (blue lines in Fig. 3.5). Both sets of results are generated using the same values of the circuit parameters. The experimental observations and the simulation results are drawn on the same figure using the same scale to demonstrate the validity of our proposed circuit model. To plot the hysteresis loop, the current was sensed using an instrumentation amplifier with its differential inputs connected across the resistor R in Fig. 3.4. Figure 3.5 reveals that the experimental results closely match the SPICE simulations.

Both experimental and simulation results demonstrate that our proposed emulator circuit of Fig. 3.4 shows nonlinearity, and a hysteresis loop appears in the I-V plane for a particular frequency range. We also observed that as we increased the frequency of the input signal, the nonlinearity gradually decreased. At a certain point,

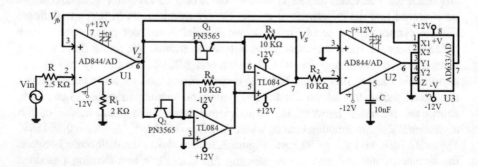

Fig. 3.4 Implemented circuit of the emulator for the current-controlled STBM memristor model

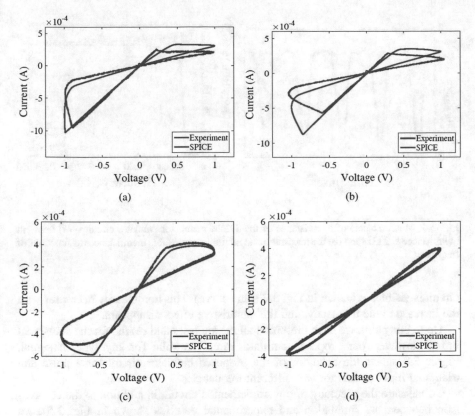

Fig. 3.5 Experimental results and SPICE simulations showing the current-voltage relation (hysteresis loop) of the STBM memristor emulator circuit of Fig. 3.4 for four different frequencies: (**a**) 1 kHz, (**b**) 1.5 kHz, (**c**) 2 kHz, and (**d**) 7 kHz

the emulator started to behave like a resistor at a higher frequency (7 kHz and higher). The experimental data were exported from the oscilloscope and redrawn using MATLAB to calculate the root mean square error (RMSE) between the experimentation and the simulation.

We also analyzed the memristance of the proposed circuit of Fig. 3.4. In Fig. 3.6a, we observed that the memristance varied with time for an applied sinusoidal signal with amplitude 1 V and frequency 2 kHz. It is observed that the memristance changed from 0.5 kΩ to 3.2 kΩ. From Fig. 3.6b, it is noticed that as we increased the frequency of the applied signal, the difference between the maximum and the minimum obtainable memristance decreased, and at some point beyond 8 kHz, the maximum and the minimum memristance coincided, while the emulator circuit acted as a pure resistor.

Figure 3.7 shows the time-domain waveforms of the input voltage and the input current of the proposed emulator circuit, where the current and the voltage have the same phase, which indicates that it is purely memristive without any series reactive element. The nonlinearity of the current in the implemented memristor emulator

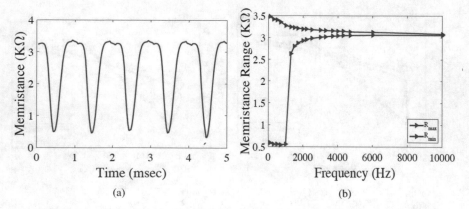

(a)　　　　　　　　　　　　　　　　　　　　(b)

Fig. 3.6 Measurements of memristance of the STBM memristor emulator circuit – (**a**) transient memristance at 2 kHz and (**b**) the maximum and minimum achievable memristances as functions of frequency

circuit is visible, as shown in Fig. 3.7 (blue curve). This nonlinearity decreases with the increase in the frequency, and the memristive effect disappears.

Up to this point, we have presented all the analyses and results for the sinusoidal input. However, our proposed emulator circuit is valid for any type of signal. Figure 3.8 shows the behavior of the proposed emulator circuit for a pulse and triangular input signals for two different frequencies.

To measure the accuracy of our implemented circuit, in addition to the comparison between the simulation and experimental data (as shown in Fig. 3.5), we calculated the root mean squared error (RMSE), as shown in Table 3.1, which listed the calculated errors. The error is coming from the parasitic element in the x and z terminals of the AD844 (a discrete component of the current feedback operational amplifiers (CFOA)) and the passive elements used in the circuit. The cause of the error is due to the uncertainty and the percentage of the error in the passive elements, which is around 10%. We like to mention that the experimentation was done using commercial off-the-shelf discrete circuit components available in the academic labs. Off-the-shelf discrete circuit components suffer significant deviation from its specified characteristics under any circumstances. Additionally, the DC offset (introduced by the IC of AD844) leads to some error in the calculation. We think that this level of error is acceptable if we compare it with other similar work published in well-known journals like [120]. To minimize this error, we need to use highly precise circuit elements in a very sophisticated and automated calibration environment. Even then, we will have some errors due to the non-idealities and variations of practical circuit components.

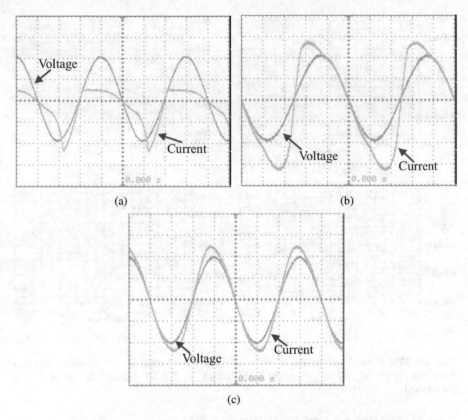

(a)   (b)

(c)

Fig. 3.7 Experimental results showing the transient input voltage $v(t)$ (yellow) and the input current $i(t)$ (blue) for the proposed emulator at (**a**) 1 kHz, (**b**) 2 kHz, and **c**) 7 kHz frequencies, respectively. The horizontal coordinate scale is 250 μs/division, and the vertical coordinate scales are CH1($v(t)$), 1 V/division, and CH2($i(t)$), 2 V/division

3.3   Implementation of an Emulator Circuit for the ThrEshold Adaptive Memristor Model (TEAM) Model

In the TEAM model, the rate of change of the memristance is a power function with power $\alpha = 9$ [88]. The power relation can be realized by the multiplication operation. We propose to utilize a voltage multiplier circuit, as shown in Fig. 3.9a, where $\alpha = 3$. If we implement the voltage multiplier circuit using AD633 as in [127], we obtain the circuit of the shaping function for the TEAM model, as shown in Fig. 3.9b. The input-output relation of the TEAM shaping function is given by Eq. (3.14), where Vo is the output voltage, Vin is the input voltage, and α is the multiplier constant.

$$V_o = \alpha^2 V_{in}^3 \qquad (3.14)$$

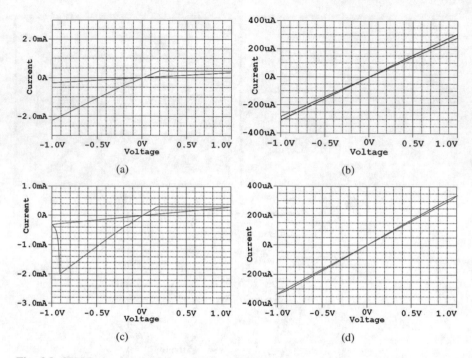

Fig. 3.8 SPICE simulation of the behavior of the proposed emulator circuit with pulse and triangular inputs: (**a**) 200 Hz(pulse), (**b**) 1 kHz(pulse), (**c**) 200 Hz (triangular), and (**d**) 2 kHz (triangular)

Table 3.1 Root mean square error (RMSE) for the STBM model

Frequency	1KHz	1.5KHz	2KHz	7KHz
RMSE	6.859e-5	13.3e-5	3.599 e-5	7.091e-6

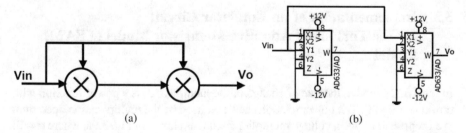

Fig. 3.9 Implementation of the proposed shaping function for the TEAM model: (**a**) block diagram and (**b**) circuit realization

Now, if we use the proposed shaping function circuit of Fig. 3.9b in the emulator circuit model of Fig. 3.1, we obtain the emulator circuit to implement the TEAM model, as shown in Fig. 3.10. The proposed circuit is realized and implemented with commercial current feedback operational amplifiers, AD844 (for U1 and U2), which

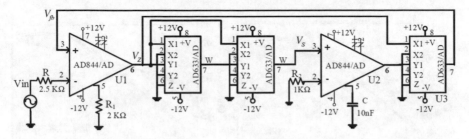

Fig. 3.10 Schematic diagram of the emulator circuit implemented for the current-controlled TEAM memristor model

are used as the second-generation current conveyors (CCII+), one analog multiplier, AD633 (U3), and some passive elements. The values of the passive elements are $R = 2.5k\Omega$, $R1 = 2k\Omega$, $R2 = 1k\Omega$, and $C = 10nF$. For simplicity, in our experimental work, we used only two voltage multipliers in a series, and the reason was to minimize the parasitic effect that was coming from the discrete components. The advantages of this design choice are higher stability of the proposed emulator circuit and suitability for several practical applications.

To measure the accuracy of the proposed emulator, we compared the numerical results obtained from the behavioral model of the TEAM model with the results obtained from the SPICE simulation. The parameters used for the behavioral model are $\alpha off = 3$, $\alpha on = 3$, $Ron = 0.9K\Omega$, $Roff = 1K\Omega$, and $k = 1e13$ for different frequencies. It can be observed from Fig. 3.11 that the SPICE simulation results and the numerical results from the behavior model match well. The root mean square error, RMSE, is less than 1.75e-3.

For the verification of the proposed emulator circuit for the TEAM model, we first implemented the shaping circuit of Fig. 3.9b using an input voltage of amplitude 10 V at 10 Hz. For the emulator circuit, we used an input voltage of amplitude 2.0 Vpp at different frequencies with a DC supply of ± 12 V. Figure 3.12a shows the nonlinear curves of the proposed shaping function obtained from the SPICE simulation and the experimental measurements conducted on the implemented shaping function circuit. It can also be observed that the simulation results and the experimental observations match well for our proposed model and circuit. Figure 3.12b shows the input-output relation of the shaping function circuit. It is evident that the relationship is asymmetric.

The hysteresis loops obtained from the simulation and the experimental data for the proposed emulator circuit of Fig. 3.10 are shown in Fig. 3.13. We observed that the emulator had a pinched hysteresis loop in the I-V plane, as expected. For instance, the result at 100 Hz shows behavior similar to the TEAM model [88] and shows significant nonlinearity in the I-V plane. With the increase of frequency, the lobe area shrinks gradually. Again, it can be observed that the experimental results match very well with the simulation results. The current was sensed using an instrumentation amplifier sensing the differential voltage across the resistance R.

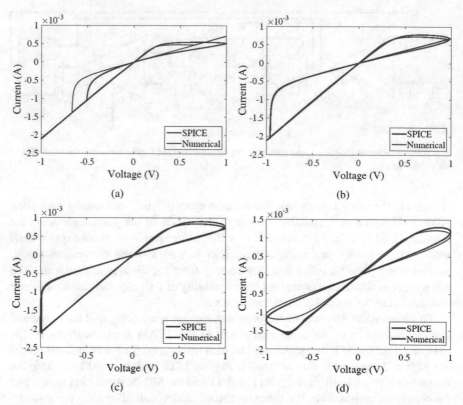

Fig. 3.11 Comparative illustrations of the SPICE simulation results obtained from the implemented emulator circuit and the numerical analysis based on the behavioral model of TEAM model showing the pinched hysteresis loop at different frequencies: (**a**) 100 Hz, (**b**) 600 Hz, (**c**) 1 kHz, and (**d**) 6.5 kHz

We also computed the RMSE between the SPICE simulation and experimental data of Fig. 3.13. Table 3.2 presents the error calculation data.

Since the memristor is a resistive element, no phase shift should be observed between the voltage and the current in the time domain. From Fig. 3.14, it appears that the proposed emulator circuit of Fig. 3.10 (based on the TEAM model) does not have a phase shift between the current and the voltage in the time domain. The current is zero whenever the voltage is zero, which is a signature property of the memristor [2, 5]. Therefore, the proposed circuit of Fig. 3.10 behaves purely like a memristive element without any series reactance. However, a closer look at Fig. 3.13 reveals that at a higher frequency, there is a minor phase shift. This is because the test circuit for the emulator is implemented with discrete components. It would be impossible to ensure an ideal zero phase shift property at a higher frequency using off-the-shelf discrete components due to their inherent parasitic effects. Therefore, the minor phase shift in Fig. 3.13d is not due to the modeling error; it is due to the

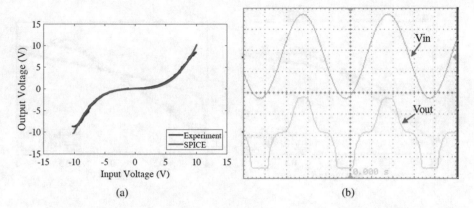

Fig. 3.12 (**a**) Experimental results and SPICE simulations showing the nonlinear behavior of the proposed shaping function for the TEAM model and (**b**) the transient input (yellow) and output voltages (blue). The horizontal coordinate scale is 250 μs/division, and the vertical coordinate scales are CH1 (V_{in}): 1 V/division, CH2 (V_o): 2 V/division

parasitic effects and imperfections of the circuit components. The input CCII ($U\,1$) has finite input resistance and capacitance. The input resistance R of our proposed circuit model (Fig. 3.1) will be affected by the input resistance of $U1$. The capacitive effect of CCII becomes prominent at higher frequencies. As discussed in [129, 140], the existence of a reactive element in series with a memristor causes a phase shift in the hysteresis. Thus, a phase shift at a higher frequency is inevitable for off-the-shelf components.

From the SPICE simulation of the proposed circuit of Fig. 3.10, we observed (see Fig. 3.15a) that the memristance changes from 0.7 $k\Omega$ to 1.8 $k\Omega$ with time for a sinusoidal signal of frequency, 1 kHz and amplitude, 1 V. Figure 3.15a shows the maximum and the minimum achievable memristance, which changes with the frequency. Here, *Rmin* is almost constant, and it represents R in Fig. 3.10. *Rmax* decreases gradually with the increase of frequency of the sinusoidal input signal, and at one point, it coincides with *Rmin*. This indicates that beyond a specific frequency, the emulator circuit starts to behave like a linear resistor.

3.4   Memristor-Based Wien Bridge Oscillator

Recently, it was claimed in different publications that passive resistors could be replaced by memristors in many applications like relaxation oscillator and Wien Bridge oscillator. Several memristor-based oscillators were illustrated in [37]. We implemented a Wien Bridge Oscillator using memristor in place of the resistor and observed that it is possible to get sustained oscillation when the poles were oscillating due to the memristive properties. To investigate the applicability of our proposed

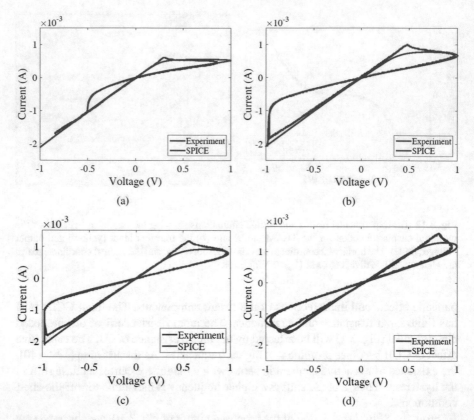

Fig. 3.13 Experimental results and SPICE simulations showing the pinched hysteresis loop in the I-V relation of the proposed emulator circuit for the TEAM model at different frequencies: (a) 100 Hz, (b) 600 Hz, (c) 1 kHz, and (d) 6.5 kHz

Table 3.2 Root mean square error (RMSE) for the TEAM model

Frequency	100 Hz	600 Hz	1KHz	6.5KHz
RMSE	4.1209e-5	6.87 e-5	7.817e-5	8.04e-5

emulator circuits, we have demonstrated the Wien Bridge oscillator implementation using the emulator for the Simmons model, as shown in Fig. 3.16a, where the resistor $R4$ has been replaced by the proposed emulator.

The oscillator is experimentally verified using the STBM emulator circuit presented in Sect. 3.2 with the following values: $R1 = 3.3K\Omega$, $R2 = 37.5K\Omega$, $R3 = 12.88K\Omega$, $C1 = C2 = 100nF$ and the DC voltage $= \pm 12\ V$. The gain of the oscillator is adjusted by the ratio between $R2$ and $R3$ to obtain a sustained sinusoidal signal, as shown in Fig. 3.16b. It is important to note that to use the memristor, the output frequency of the oscillator should be chosen within the operating range of the memristor. However, if the required output frequency is outside the range of the

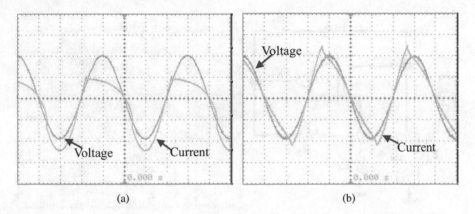

Fig. 3.14 Experimental results of the transient input voltage $v(t)$ (yellow) and the input current $i(t)$ (blue) for the proposed memristor emulator at (**a**) 100 Hz and (**b**) 6.5 kHz. The horizontal coordinate scale is 500 μs/division, and the vertical coordinate scales are CH1, 1 V/division, and CH2, 2 V/division

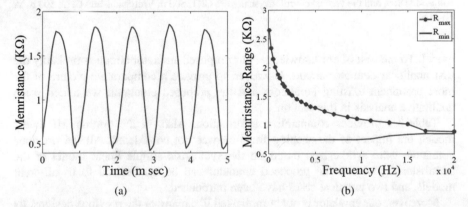

Fig. 3.15 Memristance measurement for the proposed emulator for the TEAM model: (**a**) transient memristance at 1 kHz and (**b**) the maximum and minimum achievable memristances as functions of frequency

memristive behavior, then it will act like a resistor. In our experimental work, we adjusted the circuit to oscillate with a frequency of around 500 Hz, where the memristive behavior of our circuit exists.

3.5   Comparison with the Existing Emulators

Since our emulator is a new type of nonlinear memristor emulator, it would not be very meaningful to perform a quantitative comparison of the proposed emulator with the existing emulator circuits, which are mostly based on the original HP linear

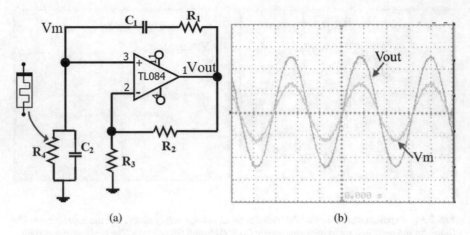

(a)   (b)

Fig. 3.16 Implementation of a Wien Bridge oscillator based on the proposed memristor emulator: (**a**) schematic diagram and (**b**) experimental results showing voltages across the memristor (blue) and the output voltage of the Wien Bridge oscillator (yellow). The horizontal coordinate scale is 500 μs/division, and the vertical coordinate scales are CH1, 500 m V/division, and CH2, 500 m V/division

model. To the best of our knowledge, our proposed emulator circuit is probably the first nonlinear emulator circuit. However, to provide a comparative picture of the most prominent existing emulators and the proposed emulator, we added some qualitative analysis in this section.

Table 3.3 provides comparative information. Also, in the original HP linear model, the impact of the shaping function was not considered. All the previous emulators were synthesized based on the symmetric simple linear model of the memristor. Therefore, the proposed emulator can be synthesized to fit different models, and two practical cases have been introduced.

Moreover, our emulator is not complicated like many of the previous designs. Its complexity depends on the complexity of the required model to fit. Our proposed model can be implemented with both the symmetric and the asymmetric shaping functions and, therefore, can be designed to fit both symmetric and asymmetric behaviors.

Another aspect of memristor emulator circuits is its input impedance, which is important because the memristor would be connected to other elements in an application, either as a passive or an active component. Therefore, input impedance should be included in any circuit representation of a physical device like a memristor. For instance, the input impedance of Lopez's emulator [120] is infinity, which would not add a loading effect if it is connected in any circuits. This would not be true for a physical memristor. In contrast with many of the existing emulators, the proposed circuit in this chapter has a finite input impedance.

Table 3.3 Comparative analysis of the proposed emulator and some of the existing emulator circuits

	[110]	[108]	[120]	[116]	This work
Type	Floating	Grounded	Floating	Grounded	Grounded
Complexity	Complex	Simple	Moderate	Simple	Moderate
Control variable	Current	Voltage	Voltage	Current	Current
Fitted model	Simple linear	Simple linear	Simple linear	Simple linear	Generic
Input impedance	Memristance	Memristance	Open circuit (infinity)	Memristance	Memristance
Lobes symmetry	Symmetric	Symmetric	Symmetric	Symmetric	Both
Tested in	Memristive networks	–	–	–	Wien oscillator/ memristive networks

3.6 Conclusion

This chapter presents a circuit development technique for practical memristor emulators to mimic the nonlinear behavior of the memristor. We demonstrated two different emulator circuits for an approximated version of the two popular and realistic memristor models: the Simmons Tunneling Barrier Model (STBM) and the ThrEshold Adaptive Memristor (TEAM) models using a linear window function. For STBM, we implemented both symmetric and asymmetric shaping function circuits. Our numerical analysis based on the behavioral model, the simulations using SPICE, and the experimental results match very well, which indicates that the proposed circuits can accurately imitate the behavior of a memristor and satisfy all three fingerprints of a memristor. Our emulator circuits have the potential to be used in many practical applications in the analog and digital domains. To verify the applicability and validity of the proposed emulator circuits, we demonstrated a Wien Bridge oscillator implementation with one of the proposed emulator circuits. The proposed circuits are practical and simple to design compared to many other emulator circuits. We presented emulator for grounded memristor model to demonstrate the new emulator circuit development techniques. The proposed emulator circuit can be easily modified to work as a floating memristor emulator by adding only one CCII to the circuit to convey the current to the other terminal and mirror its voltage to be added to the output of the multiplier.

Chapter 4
Simple Current-Controlled Memristor Emulators

4.1 Memristor Emulator Circuit Based on an Integrator and an Exponential Amplifier

With the introduction of the memristor design by the HP lab, there has been a surge of interest in performing different theoretical and experimental works by utilizing emulators to investigate various potential applications of the memristor. In this section, we propose a new memristor emulator circuit (as shown in Fig. 4.1) for the current-controlled memristor model. Our emulator circuit is an improvement over the one presented in [107, 108, 118, 127], where the nonlinearity property of the memristor was obtained using one LED (LA-541B) and two resistors. The emulator circuit proposed in Fig. 4.1 consists of a practical integrator and an exponential amplifier. We built the integrator using one CCII, one resistor, and one capacitor. We constructed the exponential amplifier using two Op-Amps, two diode-connected transistors, and two resistors. The exponential amplifier provides the required nonlinearity in memristive behavior [119].

4.1.1 Mathematical Analysis of the Proposed Emulator

As mentioned earlier, the integrator in the emulator of Fig. 4.1 consists of a CCII. Equation (4.1) exhibits the general ideal characteristics of a CCII.

$$V_Y(t) = V_X(t) \text{ and } i_Z(t) = i_X(t) \tag{4.1}$$

In the integrator circuit of Fig. 4.1, the input current (i_{in}) is created by subtracting the feedback voltage (V_{fb}) from the input voltage (V_{in}), and then this current is imposed in the capacitor. The resulting voltage across the capacitor is given by Eq. (4.2).

© Springer Nature Switzerland AG 2021
A. G. Alharbi, M. H. Chowdhury, *Memristor Emulator Circuits*,
https://doi.org/10.1007/978-3-030-51882-0_4

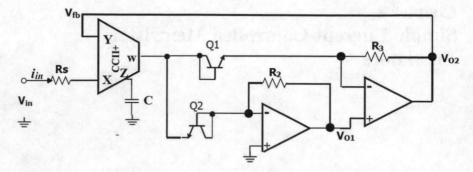

Fig. 4.1 The proposed emulator circuit based on an integrator (CCII) and an exponential amplifier

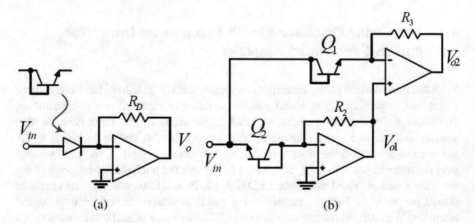

Fig. 4.2 Exponential amplifier implementation: (**a**) single-sided and (**b**) double-sided

$$V_c = \frac{1}{C} \int_0^t i_{in}(\tau)\, d\tau = \frac{q_{in}}{C} = \frac{1}{CR_s} \int_0^t \left(V_{in} - V_{fb} \right) d\tau \qquad (4.2)$$

A single-sided exponential amplifier can be built by using one Op-Amp, one resistor, and one diode, as shown in Fig. 4.2a. The output voltage is exponentially proportional to the positive input voltage and zero for the negative input voltage. The input-output relation of the exponential amplifier is shown in Eq. (4.3), where I_{ES} is the reverse saturation current, which is in order of $10^{-13}A$ and Vth is the threshold voltage.

$$V_o = I_{ES}R_D \left(1 - e^{\frac{Vin}{V_T}} \right) \approx -I_{ES}R_D e^{\frac{Vin}{V_T}} \qquad (4.3)$$

For designing a symmetric nonlinear circuit, two single-sided exponential amplifiers and a couple of reversed connected diodes are used, as shown in Fig. 4.2b. The output voltage of the first Op-Amp, V_{o1}, is shown in Eq. (4.4).

$$V_{o1} \approx I_{ES}R_2 e^{\frac{-V_{in}}{V_T}} \tag{4.4}$$

The output of the second Op-Amp, V_{o2}, is shown in Eq. (4.5).

$$V_{o2} \approx -I_{ES}\left(R_3 e^{\frac{V_{in}}{V_T}} - R_2 e^{\frac{-V_{in}}{V_T}}\right) \tag{4.5}$$

To achieve symmetry in the nonlinear circuit, we made $R_2 = R_3 = R_D$, and therefore the input-output relation of the double-sided exponential amplifier can be given by Eq. (4.6).

$$V_{o2} \approx -2I_{ES}R_D \sinh\left(\frac{V_{in}}{V_T}\right) \tag{4.6}$$

From the circuit of Fig. 4.1, it can be observed that the voltage across the capacitor of the integrator is applied to the exponential amplifier, which creates the feedback voltage, V_{fb}. By substituting Eq. (4.2) into Eq. (4.6), the feedback voltage can be found as in Eq. (4.7).

$$V_{fb} \approx -2I_{ES}R_D \sinh\left(\frac{q_{in}}{CV_T}\right) \tag{4.7}$$

The input voltage is a function of the feedback voltage and the input current, which can be expressed as $V_{in} = V_{fb} + i_{in}R_s$. Therefore, from Eq. (4.7), we can derive the expression of the input voltage as in Eq. (4.8).

$$V_{in} = i_{in}R_s - 2I_{ES}R_D \sinh\left(\frac{q_{in}}{CV_T}\right) \tag{4.8}$$

The memristance, $R_m = V(t)/i(t)$, can be given by Eq. (4.9).

$$R_m = R_s - \frac{2I_{ES}R_D}{i_{in}} \sinh\left(\frac{q_{in}}{CV_T}\right) \tag{4.9}$$

The sinh function can be the first-order approximation as $\sinh(x) = x + \ldots$. Based on this approximation, the memristance of Eq. (4.9) can be simplified to Eq. (4.10), which reveals that the memristance (R_m) is a function of the total input current and the charge $q(t)$.

$$R_m = R_s - \frac{2I_{ES}R_D}{CV_T}\left(\frac{q_{in}}{i_{in}}\right) \tag{4.10}$$

4.1.2 Circuit Realization

The proposed design of Fig. 4.1 has been simulated using SPICE and implemented using off-the-shelf components for experimental validation. The integrator is implemented using an AD844AN circuit model as the second-generation current conveyor (CCII). We used the commercially available AD844 current feedback operational amplifiers (CFOA) [138] as the CCII in the experiment. The passive elements used in the integrator are $R_s = 1.5K\Omega$ and $C = 10\ nF$. The exponential amplifier is implemented using two conventional Op-Amps (implemented using TL084), two transistors (implemented using PN2222A), and two resistors (each one of value $1K\Omega$), as shown in Fig. 4.3. We used DC supply voltages $= \pm15\ V$ in this circuit. The amplitude of the applied voltage was 2 V.

4.1.3 SPICE Simulations and Experimental Results

Figure 4.4 shows the input-output relation of the double-sided exponential amplifier of the proposed design. It is observed that the nonlinear curve is symmetric, which indicates that the resulting memristor emulator would also be symmetric. If needed, an asymmetric memristor emulator can be designed by using two different transistors in the double-sided exponential amplifier, for example, the designers can use PNP and NPN transistors.

The simulation results shown in Fig. 4.5 demonstrate that the proposed circuit exhibits the fingerprints of the memristors that were discussed in detail in [13]. The I-V curve shows the pinched hysteresis behavior and crosses at the origin. The hysteresis lobe area shrinks with the increase of the frequency, where the nonlinearity of the memristor decreases gradually. At a certain point, the hysteresis loop tends to become a straight line when the applied frequency is increased, and the emulator starts to behave like a resistor.

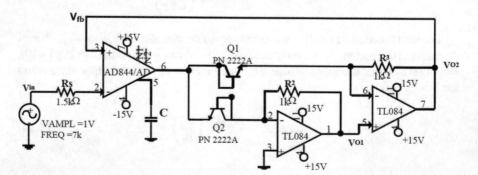

Fig. 4.3 The schematic diagram of the proposed memristor emulator

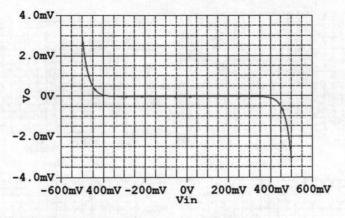

Fig. 4.4 SPICE simulation of the double-sided exponential amplifier

Figure 4.5a–c show a perfect pinched hysteresis loop for the frequency range from 3 *kHz* to 10 *kHz*. However, from Fig. 4.5d, we observed that the pinched hysteresis loop collapsed at a higher frequency of 15 *kHz* compared to the loops in Fig. 4.5a, b, c. If we continue to increase the frequency, we notice a distinct change at 20 *kHz*. At this point, there is no pinched hysteresis loop anymore, and the emulator acts as a pure resistor. This behavior validates the unique property of a memristor, known as the frequency dependence of the hysteresis behavior. It is also observed that the hysteresis loop is restricted in the first and third quadrants. Hence, we can say that our proposed circuit represents a passive element that satisfies Chua's passivity condition [5]. These results demonstrate that the proposed emulator can provide the unique features of the memristor.

The emulator proposed in Fig. 4.1 is implemented in the lab for the experimental measurement using the circuit components and values mentioned in Sect. 4.1.2. The amplitude of the applied voltage was 2 V. To draw the I-V curves, the current was sensed using an instrumentation amplifier, sensing the differential voltage across the resistance R_S with a gain of (1 + 50 k/300). The experimental results are shown in Fig. 4.6. All the components were off-the-shelf and readily available in the market. It is worth mentioning that the input current was not ideally integrated since we used a resistor R_I of value 2 $k\Omega$ in parallel with the capacitor C. Our circuit is still valid even though the integrator is not ideal.

Also, from the experimental measurements of Fig. 4.6, it is observed that the proposed emulator circuit can mimic the behavior of the memristor. As in the simulation results of Fig. 4.5, we also notice the signature pinched hysteresis loop at a lower frequency in the I-V plane in Fig. 4.6. At higher frequency, the lobe area starts to shrink, and the emulator circuit tends to behave as a pure resistor. Therefore, the experimental results (Fig. 4.6) and the simulation (Fig. 4.5) match and validate that the proposed emulator ensures one of the three signatures of a memristor, as illustrated in [13].

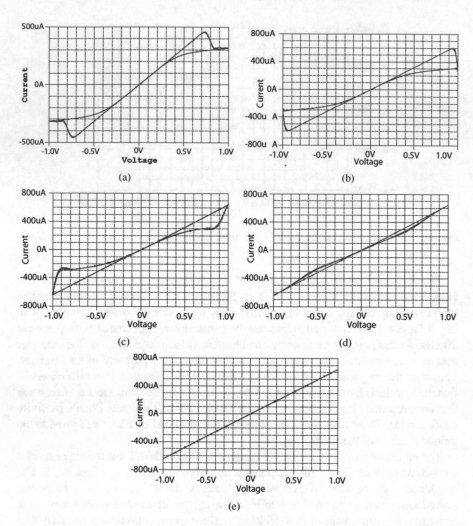

Fig. 4.5 Simulation of the I-V characteristics of the proposed emulator of Fig. 4.3 at various frequencies – (**a**) 3 kHz, (**b**) 7 kHz, (**c**) 10 kHz, (**d**) 15 kHz, and (**e**) 20 kHz

It is well-known that the memristor is a resistive element, so there is no phase shift between the current and the voltage, which is clear in the proposed emulator, as shown in Fig. 4.7a. The current is zero whenever the voltage is zero, which is another signature of a memristor. If a phase shift exists, this means that there is a reactive element attached to the device. This implies that the proposed memristor emulator circuit is resistive without any reactive element attached. Figure 4.7b shows the change in the memristance due to the applied sinusoidal signal, where the memristance changes from 1.6 $k\Omega$ to 3.45 $k\Omega$.

The maximum and the minimum achievable memristance is calculated for the proposed emulator circuit using SPICE by sweeping the frequency. Figure 4.8,

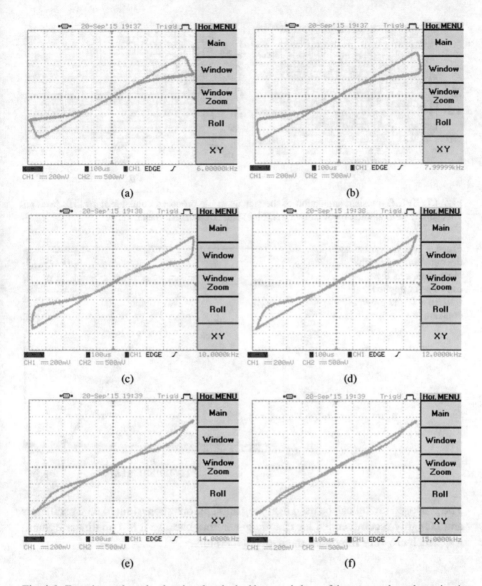

Fig. 4.6 Experimental results showing the pinched hysteresis loop of the proposed emulator circuit of Fig. 4.3 at various frequencies – (**a**) 6 kHz, (**b**) 8 kHz, (**c**) 10 kHz, (**d**) 12 kHz, (**e**) 14 kHz, and (**f**) 15 kHz

which provides the results of the maximum and the minimum achievable memristance, shows that the gap decreases gradually with the increase of frequency. The memristance gap keeps decreasing until it reaches zero at very high frequencies, where the emulator acts like a linear resistor. At low frequencies, the maximum and the minimum achievable memristance tends to saturate at specific values due to the power supply limitation of the circuit, which is well-known in all emulation circuits.

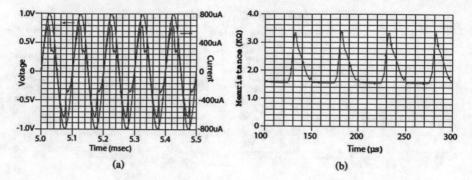

Fig. 4.7 SPICE transient simulation of the signals in the proposed emulator at 10 kHz (**a**) input voltage *v(t)* (red) and the input current *i(t)* (blue) and (**b**) transient memristance

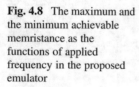

Fig. 4.8 The maximum and the minimum achievable memristance as the functions of applied frequency in the proposed emulator

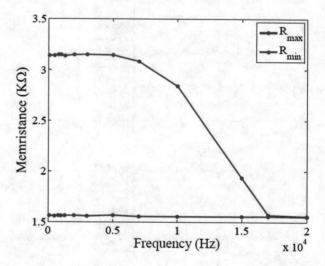

The proposed circuit can be used in analog applications such as sinusoidal oscillators and is highly recommended for chaotic generators because it has nonlinear dynamics, which are very important in chaotic generators.

4.2 A New Simple Emulator Circuit for Current-Controlled Memristor

In an attempt to further simplify the emulator design for the current-controlled memristor, a new design has been proposed in this section. The proposed emulator circuit is a modification and improved version of the original emulator circuit presented in [119, 127]. The new simpler emulator circuit is shown in Fig. 4.9, which still consists of a practical integrator and an exponential amplifier (similar to

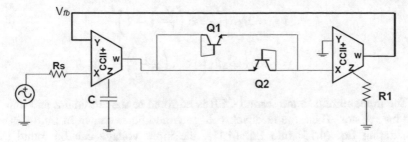

Fig. 4.9 The proposed simple current-controlled memristor emulator circuit

the one proposed in Fig. 4.1). The practical integrator is implemented using one CCII. However, the change/difference in the designs of Fig. 4.1 and Fig. 4.9 is the implementation of the exponential amplifier. In Fig. 4.9, the exponential amplifier is also implemented using one CCII along with two diode-connected transistors and one resistor. The input current is integrated using the integrator that represents the charge, which is applied to the exponential amplifier to generate the feedback voltage to change the state of the circuit. The new design of Fig. 4.9 is more straightforward because both of the critical components of the emulator are implemented using CCII, and the overall count of the circuit components is lower.

4.2.1 Mathematical Analysis of the Proposed Emulator

The input current of the circuit is a function of the applied voltage (V_{in}) and the feedback voltage (V_{fb}), as shown in Eq. (4.11). The input current is injected into the first capacitor C that generates the voltage V_c, as shown in Eq. (4.12). The current passing through the diode-connected transistors is given by Eq. (4.13).

$$i_{in} = \frac{V_{in} - V_{fb}}{R_s} \tag{4.11}$$

$$V_c = \frac{1}{C} \int_0^t i_{in}(\tau)\, d\tau \tag{4.12}$$

$$I_Q = I_S\left(e^{\frac{V_{be}}{V_{th}}} - 1\right) \tag{4.13}$$

Here V_{th} is the threshold voltage, and I_s is the saturation current. The total input current in the terminal X of the second CCII can be given by Eq. (4.14).

$$I_x = I_S\left(e^{\frac{V_C}{V_{th}}} + e^{\frac{-V_C}{V_{th}}} - 2\right) = 2I_S\left(\cosh\left(\frac{V_C}{V_{th}}\right) - 1\right) \tag{4.14}$$

$$V_{fb} = 2I_S R_1 \left(\cosh \left(\frac{V_C}{V_{th}} \right) - 1 \right) \tag{4.15}$$

$$V_{in} = i_{in} R_s + 2I_S R_1 \left(\cosh \left(\frac{q}{CV_{th}} \right) - 1 \right) \tag{4.16}$$

The input current to the second CCII is mirrored to the terminal Z and imposed into the resistor. Then, the feedback voltage would be as shown in Eq. (4.15). By substituting Eq. (4.15) into Eq. (4.11), the input voltage can be found as in Eq. (4.16). The resulting memristance $(R_m = V(t)/i(t))$ can be given by Eq. (4.17).

$$R_m = R_s + 2 \frac{I_S R_1}{i_{in}} \left(\cosh \left(\frac{q}{CV_{th}} \right) - 1 \right) \tag{4.17}$$

It is worth noticing that the memristance R_m is a function of the total current and the charge $q(t)$. Based on the expansion of *cosh*, the memristance can be reduced to Eq. (4.18).

$$R_m \approx R_s + \frac{I_S R_1}{C^2 V_{th}^2} \frac{q(t)^2}{i_{in}} \tag{4.18}$$

4.2.2 Circuit Realization

The proposed circuit is implemented using two AD844AN as CCII for the integrator and the exponential amplifier. As in the previous experiment, we used commercial AD844 CFOA as CCII and two transistors (PN2222A). The passive elements used are $R_S = 1.1 k\Omega$, $R_1 = 1 k\Omega$ and $C = 0.1\ \mu F$. The implemented circuit is shown in Fig. 4.10. We used DC supply voltages $= \pm 15\ V$. The applied voltage signal has an amplitude of 2 V.

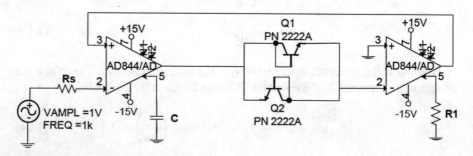

Fig. 4.10 The schematic diagram of the simple current-controlled memristor emulator

4.2.3   SPICE Simulations and Experimental Results

The simulation results of Fig. 4.11 shows that the proposed circuit exhibits the fingerprints of a real memristor, which were discussed in Chap. 1. The simulation of the proposed emulator shows nonlinearity in the (I-V) plane for a particular frequency range as expected. The nonlinearity disappears with the increase of frequency. Above a certain frequency, the emulator starts to behave like a resistor (as shown in Fig. 4.11). From Fig. 4.11, it is also noticed that at lower frequencies

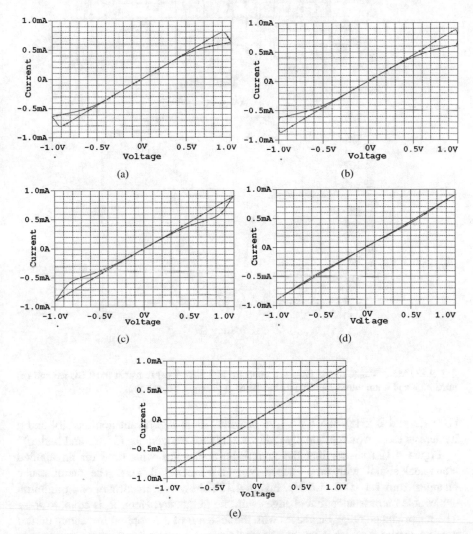

Fig. 4.11 Simulation of the I-V characteristics of the proposed emulator circuit of Fig. 4.10 at frequencies – (**a**) 700 Hz, (**b**) 1 kHz, (**c**) 1.5 kHz, (**d**) 2 kHz, and (**e**) 3 kHz

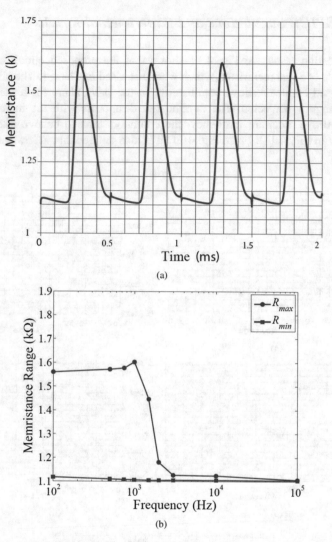

Fig. 4.12 Memristance of the proposed emulator at 1 kHz – (**a**) transient memristance and (**b**) maximum and minimum achievable memristance at different frequencies

(700 Hz and 2 kHz), the I-V curve provides more significant nonlinearity and a hysteresis loop, which gradually reduces at higher frequencies (3 kHz and higher).

Figure 4.12a shows that the memristance changes with time for an applied sinusoidal signal with amplitude 1 V and frequency 1 *k*Hz. The memristance changes from 1.1 *k*Ω to 1.6 *k*Ω. Figure 4.12b shows the maximum and minimum achievable memristance that changes with the frequency. Here, R_s is equal to R_{min}. The memristance range decreases with the increase of the applied frequency until it reaches a point at which it acts as a linear resistor. However, for frequencies less than 1 *k*Hz, the maximum and the minimum resistances are constant due to the saturation of the circuit. Therefore, we can conclude that the proposed emulator circuit can accurately imitate the characteristics of the memristor within a particular range of frequencies.

The proposed memristor circuit of Fig. 4.10 is also implemented in the lab with the components and signal values mentioned in Sect. 4.2.2. To plot the I-V curves, the current was sensed using an instrumentation amplifier, which measured the differential voltage across the resistance R_s with a gain of $(1 + 50 \text{ k}/300)$. The tested circuit in the lab gives the same results as in the SPICE simulations. It is worth noting that by reducing the value of C, we can still get the pinched hysteresis loop of the memristor at higher frequencies. This is a well-known property for all the memristor emulators [127].

Figure 4.13 shows the effect of changing frequency from 700 Hz to 3 kHz. The hysteresis loops decrease with the increasing frequency. Figure 4.14 shows the time-domain curves of the voltage and the current of the memristor. It is observed that the current and the voltage have the same phase, which means that it is purely resistive without any series reactive element.

4.3 Memristor Emulator Based on a Single CCII

In this section, the third and the simplest emulator circuit (compared to the two proposed in Sects. 4.1 and 4.2) is illustrated. The proposed third emulator circuit for the current-controlled memristor is shown in Fig. 4.15, which is a modified and improved version of the two emulator circuits presented in Sects. 4.1 and 4.2. In this new design, the number of circuit components is further reduced. This is one of the simplest emulators for the current-controlled memristor available in the literature.

4.3.1 Mathematical Analysis of the Proposed Emulator

The current passing through the upper and lower diode-connected transistors (Q_1 and Q_2 in Fig. 4.15) are I_1 and I_2, respectively, and can be given by Eq. (4.19).

$$I_1 = I_{ES}\left(e^{\frac{V_x - V_{fb}}{V_T}} - 1\right), I_2 = I_{ES}\left(e^{-\frac{V_x - V_{fb}}{V_T}} - 1\right) \tag{4.19}$$

The total current ($I_1 + I_2$) passes through R_1 and generates the feedback voltage V_{fb}, which is given by Eq. (4.20).

$$V_{fb} = (I_1 + I_2)R_1 = 2I_{ES}\left(R_1 \cosh\left(\frac{V_x - V_{fb}}{V_T}\right) - 1\right) \tag{4.20}$$

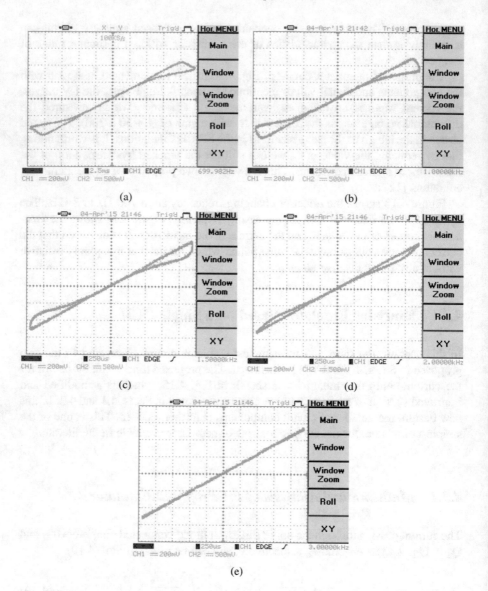

Fig. 4.13 Experimental results showing the pinched hysteresis loop of the proposed emulator circuit of Fig. 4.10 at various frequencies – (**a**) 700 Hz, (**b**) 1 kHz, (**c**) 1.5 kHz, (**d**) 2 kHz, and (**e**) 3 kHz

It is observed that Eq. (4.20) is an implicit relation of V_{fb}. To separate the variables and obtain an explicit relation of V_{fb}, the second-order approximation of cosh is used, and the resulting expression of V_{fb} is as shown in Eq. (4.21).

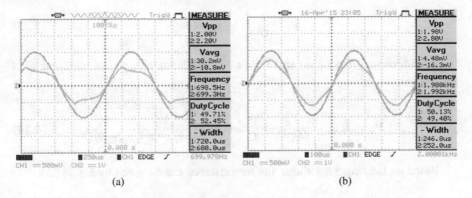

(a)   (b)

Fig. 4.14 Time-domain behaviors of the input voltage $v(t)$ (yellow) and the input current $i(t)$ (blue) of the proposed memristor emulator of Fig. 4.10 at frequencies – (**a**) 700 Hz and (**b**) 2 kHz

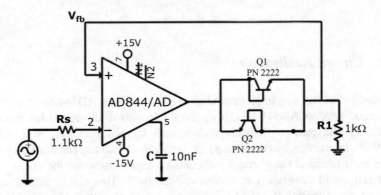

Fig. 4.15 The schematic diagram of the proposed single CCII-based emulator

$$V_{fb} = 2I_{ES}R_1 \left(1 + \frac{V_x^2}{V_T^2} - \frac{2V_x V_{fb}}{V_T^2} + \frac{V_{fb}^2}{V_T^2} \right) \quad (4.21)$$

Solving the quadratic equation of (4.21) results in Eq. (4.22).

$$V_{fb} = \frac{V_T^2}{4I_{ES}R_1} + V_1 \pm 0.5V_T \sqrt{\frac{V_T^2}{4I_{ES}R_1} + \frac{2V_x}{I_{ES}R_1}} \quad (4.22)$$

The input current is injected into the first capacitor C to generate V_C as in Eq. (4.23).

$$V_x = \frac{1}{C} \int_0^t i_{in}(\tau) \, d\tau = \frac{q}{C} \tag{4.23}$$

By substituting Eq. (4.23) into Eq. (4.22), the input voltage can be found as in Eq. (4.24).

$$V_{in} = i_{in}R_s + \frac{V_T^2}{4I_{ES}R_1} + \frac{q}{C} \pm 0.5V_T \sqrt{\frac{V_T^2}{4I_{ES}^2R_1^2} + \frac{2q}{I_{ES}R_1C}} \tag{4.24}$$

Based on the above equations, the memristance can be given by Eq. (4.25).

$$R_m = R_s + \frac{1}{i_{in}} \left(\frac{V_T^2}{4I_{ES}R_1} + \frac{q}{C} \pm 0.5V_T \sqrt{\frac{V_T^2}{4I_{ES}^2R_1^2} + \frac{2q}{I_{ES}R_1C}} \right) \tag{4.25}$$

4.3.2 Circuit Realization

The proposed circuit was implemented as in Fig. 4.15 with AD844AN (we used the commercial version AD844 CFOA as the CCI), two diode-connected transistors (PN2222A) and passive elements with values $R_s = 1.1 \ k\Omega$, $R_1 = 1 \ k\Omega$, and $C = 10 \ nF$. In practice, by connecting any capacitor C with any resistor R in parallel, we can get a nonideal integrator. By choosing a higher value for the resistor R, the circuit can avoid integration at a zero voltage drop. This circuit acts as a valid integrator, even though it would behave as a nonideal circuit. Here, a resistor of value 900 Ω is used in parallel to the capacitor to implement a nonideal integrator. It is a trade-off for the simplicity of the circuit.

4.3.3 SPICE Simulations and Experimental Results

Figure 4.16 shows that this simplest possible emulator could mimic the behavior of a memristor. It is also observed that the hysteresis loop is restricted in the first and the third quadrants with symmetrical behavior. Therefore, we can conclude that the proposed emulator of Fig. 4.15 acts as a passive element, which satisfies Chua's condition in [2, 5, 10]. The lobe area of the pinched hysteresis loop is larger at lower frequencies and disappears at high frequencies. The pinched hysteresis loop

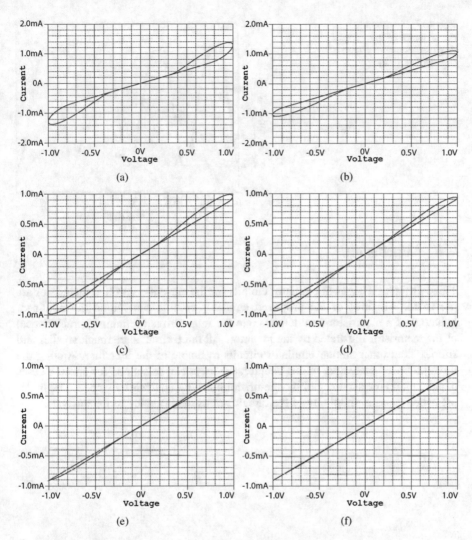

Fig. 4.16 Simulation of the I-V characteristics of the proposed emulator circuit of Fig. 4.15 at frequencies – (**a**) 3 kHz, (**b**) 7 kHz, (**c**) 10 kHz, (**d**) 12 kHz, (**e**) 15 kHz, and (**f**) 20 kHz

becomes a straight line above a certain frequency. This behavior validates the unique property of a memristor, which is known as the frequency dependence of pinched hysteresis loops. Figure 4.17 illustrates that the memristance varies with time for the applied sinusoidal signal with amplitude 1 V and frequency 10 *k*Hz. The memristance changes from 0.7 *k*Ω up to 1.12 *k*Ω.

Fig. 4.17 Transient
memristance of the
proposed emulator circuit of
Fig. 4.15 at 10 kHz

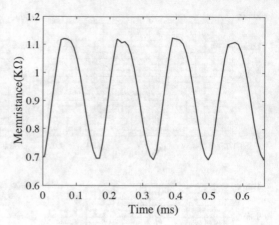

4.4 Conclusion

Three different designs of emulator circuits for the current-controlled memristor are
presented. All three of the proposed emulator circuits can mimic the nonlinear
behavior of a memristor and demonstrate the fingerprints (defining characteristics)
of the memristor illustrated by the inventor. All three circuits are much smaller and
simpler than other similar emulator circuits available in the literature. Among the
three proposed emulators in this chapter, the users can select one for specific cases
depending on the accuracy of the memristor behavior analysis they need. The trade-
off is between the simplicity of the circuit versus the accuracy of the analysis.

Chapter 5
Generic and Practical Emulators for the Voltage-Controlled Memristor Models

5.1 Emulator Circuit Design for the Grounded Voltage-Controlled Memristor Models

Most of the available emulator circuit designs are for grounded memristor models. Again, the majority of those emulators are for the current-controlled memristor models. Here, a new grounded emulator circuit design technique is introduced for the voltage-controlled memristor models. The emulator circuit design proposed in this section would fit different practical memristive models discussed in Chap. 2. The generic emulator architecture shown in Fig. 5.1 is comprised of a shaping circuit, a voltage difference circuit, a voltage integrator, and a multiplier. The shaping circuit provides the nonlinear relation of the voltage that controls the rate of change of the state variable of the memristor. The voltage difference circuit and the integrator are implemented using the second-generation current conveyor (CCII).

5.1.1 Mathematical Modeling of the Proposed Emulator

The characteristics of an ideal CCII can be given by Eq. (5.1).

$$V_Y(t) = V_X(t) \text{ and } i_X(t) = i_Z(t) \tag{5.1}$$

The input current in the circuit of Fig. 5.1 is shown in Eq. (5.2).

$$i_{in} = \frac{V_{in} - V_{fb}}{R} \tag{5.2}$$

where V_{fb} represents the feedback voltage (output of the multiplier). The input voltage is shaped using a nonlinear function, $f(V_{in})$, and then the output voltage is

© Springer Nature Switzerland AG 2021
A. G. Alharbi, M. H. Chowdhury, *Memristor Emulator Circuits*,
https://doi.org/10.1007/978-3-030-51882-0_5

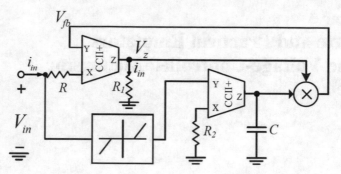

Fig. 5.1 The proposed generic emulator circuit for the grounded voltage-controlled memristor model

integrated and multiplied by the voltage (V_z) of the second CCII, which results in the feedback voltage, as is shown in Eq. (5.3).

$$V_{fb} = \frac{\alpha V_z}{R_2 C} \int_0^t f(V_{in}(\tau))\, d\tau \qquad (5.3)$$

where α is the multiplier constant, and f is the shaping function. The feedback voltage as a function of the input current is shown in Eq. (5.4).

$$V_{fb} = \frac{\alpha i_{in} R_1}{R_2 C} \int_0^t f(V_{in}(\tau))\, d\tau \qquad (5.4)$$

By substituting Eq. (5.4) into Eq. (5.2), the voltage-current relation can be found as in Eq. (5.5).

$$V_{in} = \left(R + \frac{\alpha R_1}{R_2 C} \int_0^t f(V_{in}(\tau))\, d\tau\right) i(t) \qquad (5.5)$$

The memristance is given by Eq. (5.6), and the rate of change in the memristance is as shown in Eq. (5.7).

$$R_m = R + \frac{\alpha R_1}{R_2 C} \int_0^t f(V_{in}(\tau))\, d\tau \qquad (5.6)$$

$$\frac{dR_m}{dt} = \frac{\alpha R1}{R_2 C} f(V_{in}(t)) \qquad (5.7)$$

It is evident from the equation that the rate of change of the memristance is proportional to the designed shaping function. Therefore, by choosing appropriate components to construct a window function, the required behavior from the memristor emulator can be obtained.

5.1.2   Realization of a Simpler Model

The model presented in Sect. 5.1.1 is generic and can be configured for different memristor models by changing the shaping circuit. The simplest possible model can be obtained by not having a shaping circuit, which would be a particular case of the model presented in Sect. 5.1.1. In this specific case, the shaping function $f(V_{in}) = V_{in}$, which represents a voltage-controlled linear HP model [3]. The voltage-current relation would be as shown in Eq. (5.8), where $\phi(t)$ represents the flux. The memristance for this particular case would be as in Eq. (5.9).

$$V_{in} = \left(R + \frac{R_1}{R_2 C} \phi(t)\right) i_{in}(t) \tag{5.8}$$

$$R_m = R + \frac{R_1}{R_2 C} \phi(t) \tag{5.9}$$

It is clear that the memristance R_m is a function of the flux $\phi(t)$. This case is validated by SPICE simulations and experimentations.

5.1.3   Circuit Realization

The proposed emulator circuit is designed using two second-generation current conveyors (CCII) (implemented using the commercial AD844 CFOA [138]), a voltage multiplier (implemented using the commercial AD633 multiplier [139]), and some passive elements of values $R = R_1 = 1\ k\Omega$, $R_2 = 10\ k\Omega$, and $C = 0.01\ \mu F$, as shown in Fig. 5.2. The DC supply voltages are ± 12 V, and the applied voltage signal has an amplitude of 2 V.

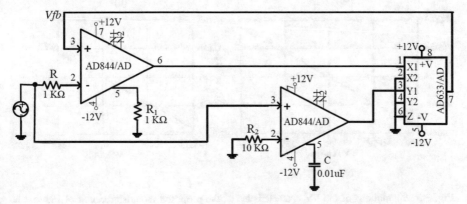

Fig. 5.2 The schematic diagram of the simple voltage-controlled memristor emulator

5.1.4   SPICE Simulations and Experimental Results

The SPICE simulations of the I-V characteristics of the implemented emulator circuit of Fig. 5.2 are shown in Fig. 5.3 for different frequencies. It is observed that the proposed emulator circuit exhibits the fingerprints (identifying characteristics) of a memristor, as introduced in [5, 13]. A memristor acts as a nonlinear device, and a pinched hysteresis loop is observed in the I-V plane for a particular range of frequency. With the increase of frequency, lobe areas of the hysteresis loop start to shrink, and at a certain point, the loop becomes a straight line, and the memristor acts like a linear resistor, as shown in Fig. 5.3.

A sinusoidal signal of 500 Hz of frequency and 1 V of amplitude is applied to observe the changes of memristance, which is illustrated in Fig. 5.4. It is noticed that the change of the memristance is also sinusoidal. The maximum and minimum values of memristance are 1.8 $k\Omega$ and 0.3 $k\Omega$, respectively. Figure 5.5 shows the maximum and minimum achievable memristance (R_{max} and R_{min}), which vary with the sweeping of the applied frequency. The difference between R_{max} and R_{min} is significant at lower frequencies (less than 1 kHz). However, as the frequency increases, the gap between achievable R_{max} and R_{min} tends to decrease. It is seen that R_{min} starts with almost 0.3 $k\Omega$ and increases drastically and then grows very

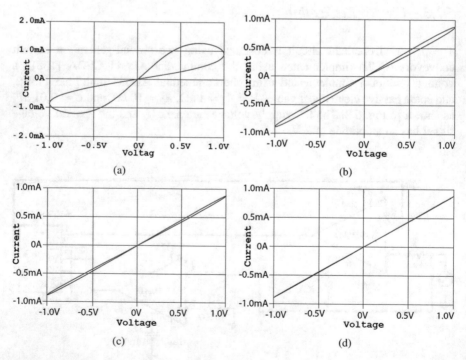

Fig. 5.3 Simulations of the I-V characteristics of the proposed memristor circuit of Fig. 5.2 at various frequencies – (**a**) 200 Hz, (**b**) 1.5 kHz, (**c**) 3 kHz, and (**d**) 5 kHz

Fig. 5.4 Transient
memristance of the emulator
circuit of Fig. 5.2 at 500 Hz

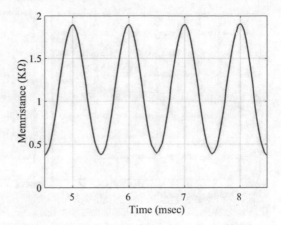

Fig. 5.5 The maximum and
the minimum achievable
memristance (R_{max} and
R_{min}) as a function of
frequency in the emulator
circuit of Fig. 5.2

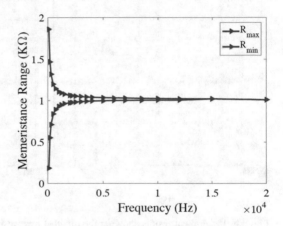

slowly for some time and merges with R_{max}. Similar behaviors are observed for R_{max} as well. It is apparent that, at a certain point (approximately 7 kHz), the memristance gap becomes zero. Therefore, at any point over this frequency, $R_{max} = R_{min}$, which indicates that the memristor emulator circuit of Fig. 5.2 acts as a pure resistor. It can be concluded that from a very low frequency until 7 kHz, the proposed emulator can accurately mimic the nonlinear property of a real memristor. For any frequency above this range, this emulator circuit acts like a linear resistor. By changing the parameters of the emulator circuit, these ranges can be varied.

For the experimental measurements, the circuit of Fig. 5.2 is implemented using the components mentioned in Sect. 5.1.3. To plot the I-V curves, the current is measured using an instrumentation amplifier sensing the differential voltage across the resistance R. The experimental observations (see Fig. 5.6) are in agreement with the SPICE simulations (see Fig. 5.5). From Fig. 5.7, which shows the waveform of the input voltage and the input current of the proposed emulator circuit, it is observed that there is no phase shift between the current and voltage. It is also noticed that there is no current when the voltage is zero, which validates a significant property of

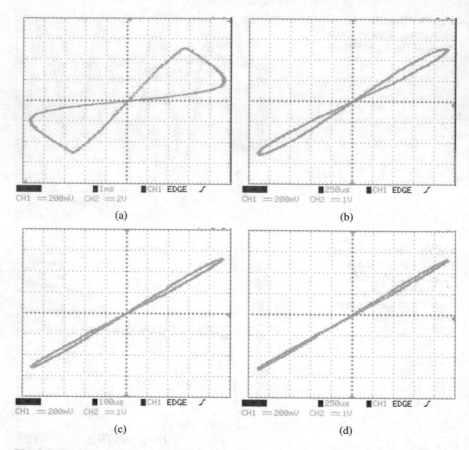

Fig. 5.6 Experiment results showing the pinched hysteresis loop in the I-V characteristics of the implemented emulator circuit of Fig. 5.2 at different frequencies (**a**) 200 Hz, (**b**) 1.5 kHz, (**c**) 3 kHz, and (**d**) 5 kHz

the memristor [5]. Therefore, we conclude that the proposed emulator circuit is purely resistive without having any reactive element attached in series.

5.2   Emulator Circuit Design for the Floating Voltage-Controlled Memristor Models

In this section, an emulator circuit design is presented for the floating memristor models. The proposed circuit of Fig. 5.8 is comprised of a shaping circuit, a voltage difference circuit, a voltage integrator, and a multiplier. The shaping circuit provides the nonlinear voltage characteristics that control the rate of change of the state variable of the memristor. The voltage difference circuit and the integrator are

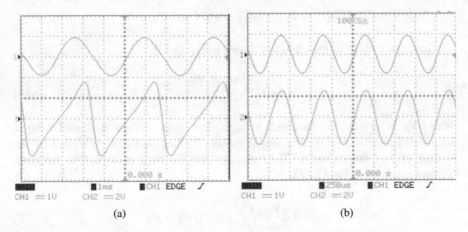

Fig. 5.7 Input voltage $v(t)$ (yellow) and input current $i(t)$ (blue) waveforms of the implemented emulator circuit of Fig. 5.2 at (**a**) 300 Hz and (**b**) 2 kHz

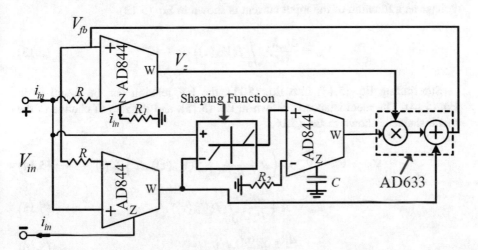

Fig. 5.8 Proposed emulator circuit design for the floating memristor models

implemented using the second-generation current conveyor (CCII) as in the emulator circuits presented in the earlier chapters.

5.2.1 Mathematical Modeling of the Proposed Emulator

The characteristics of an ideal CCII are represented in Eq. (5.10).

$$V_Y(t) = V_X(t), i_X(t) = i_Z(t), \text{and } V_W = V_Z \tag{5.10}$$

The input current of the circuit is shown in Eq. (5.11).

$$i_{in}(t) = \frac{V_{in}^+ - V_{fb}}{R} \tag{5.11}$$

where V_{fb} represents the feedback voltage. The input voltage is shaped using a nonlinear function, $f(V_{in})$. The output voltage is integrated and multiplied by the voltage (V_z) of the second CCII to generate the feedback voltage. The feedback voltage is shown in Eq. (5.12).

$$V_{fb} = \frac{\alpha V_z}{R_2 C} \int_0^t f(V_{in}(\tau)) \, d\tau + V_{in}^- \tag{5.12}$$

where α is the multiplier constant, and f is the shaping function. The feedback voltage as a function of the input current is shown in Eq. (5.13).

$$V_{fb} = \frac{\alpha i_{in} R_1}{R_2 C} \int_0^t f(V_{in}(\tau)) \, d\tau + V_{in}^- \tag{5.13}$$

Substituting Eq. (5.13) into Eq. (5.11), the I-V relation can be found as in Eq. (5.14). The memristance is shown in Eq. (5.15), and the rate of change of the memristance is shown in Eq. (5.16).

$$V_{in} = V_{in}^+ - V_{in}^- = \left(R + \frac{\alpha R_1}{R_2 C} \int_0^t f(V_{in}(\tau)) \, d\tau \right) i_{in}(t) \tag{5.14}$$

$$R_m = R + \frac{\alpha R_1}{R_2 C} \int_0^t f(V_{in}(\tau)) \, d\tau \tag{5.15}$$

$$\frac{dR_m}{dt} = \frac{\alpha R_1}{R_2 C} f(V_{in}(\tau)) \tag{5.16}$$

From Eq. (5.16), we observed that the rate of change of the memristance is proportional to the shaping function. We can develop the required memristor emulator by choosing and building an appropriate window function. The HP model is linear, as shown in Eq. (1.7) of Chap. 1, and there is no need for a shaping function. Therefore, the simplest model can be obtained without the shaping circuit, where $f(V_{in}) = V_{in}$ represents a voltage-controlled HP model [3]. The I-V relation is shown in Eq. (5.17), where $\phi(t)$ represents the flux. The memristance is given by Eq. (5.18), which is a function of the flux $\phi(t)$.

$$V_{in}(t) = \left(R + \frac{\alpha R_1}{R_2 C} \phi(t) \right) i_{in}(t) \tag{5.17}$$

$$R_m(t) = R + \frac{\alpha R_1}{R_2 C} \phi(t) \tag{5.18}$$

5.2.2 Circuits Realization

In all the emulator circuits presented in this book, the second-generation current converter (CCII) is a key component. CCIIs have been widely used in various circuit designs. In practice, researchers use the current feedback operational amplifier (CFOA) (the commercially available version is AD844) as CCII because the AD844 consists of a CCII followed by a voltage buffer [138, 141]. This arrangement makes it suitable for many applications. The floating emulator design of Fig. 5.8 is implemented in the simulation environment and experimental setup using three AD844 [138] as CCIIs, one AD633 [139] as the voltage multiplier, and several passive elements $R = R_1 = 1\ k\Omega$, $R_2 = 10\ k\Omega$, and $C = 0.01\ \mu F$. We used $\pm 9\ V$ as DC supply voltages. For the simulation, we used SPICE. For the experimental measurements, we used the Digilent Electronics Explorer Board and WaveFormsTM software. The data are exported directly to MATLAB without alterations to draw the hysteresis loop.

It is important to note that the emulator circuit of Fig. 5.8 is a generic architecture for all floating memristor models. This circuit can be used for implementing the HP model, GMM model, and VTEAM model. The differentiating component would be the shaping circuit.

5.2.3 Emulator Implementation for the HP Model

If we remove the shaping function circuit from Fig. 5.8 and implement the resulting circuit using the components listed in Sect. 5.2.2, we get the emulator circuit for the HP memristor model, as shown in Fig. 5.9. Here, we only provided the experimental measurement-based observations for the emulator circuit of Fig. 5.9.

It is observed that this emulator circuit exhibits all the fingerprints of a memristor listed in [13]. The emulator circuit acts as a nonlinear device and shows the pinched hysteresis loop in the I-V plane for a particular range of frequencies (see Fig. 5.10). It is evident that as we keep increasing the frequency, the lobe area of the hysteresis loop tends to shrink.

In the previous cases, we only used a sinusoidal input signal. In this experiment, we investigated the behavior of the proposed emulator circuit for other types of input signals as well. Figure 5.10a shows the I-V characteristics for sinusoidal input

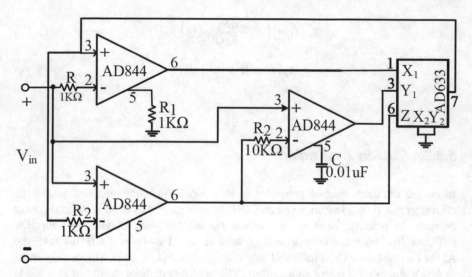

Fig. 5.9 Emulator circuit implementation for the linear HP memristor model

signals at four different frequencies, and Fig. 5.10b shows the I-V curve for the triangular input signals for two different frequencies.

Figure 5.11a shows the change of memristance in time when a sinusoidal signal of 500 Hz of frequency and 2 V of amplitude is applied. It is observed that the nature of the change of the memristance is also sinusoidal. The maximum and the minimum values of the memristance (R_{max} and R_{min}) are 1.8 $k\Omega$ and 0.4 $k\Omega$, respectively. Figure 5.11b shows that the maximum and minimum achievable memristance varies with the variation of the applied frequency. The difference between R_{max} and R_{min} is high at low frequency, and the gap decreases with the increase of frequency. R_{min} starts at 0.2 $k\Omega$ and increases drastically and then asymptotically converges towards R_{max}. The behavior of R_{max} is similar but opposite. At approximately 6 kHz, the memristance range converges to zero, which means that over a specific frequency, $R_{max} = R_{min}$, the memristor emulator circuit acts like a pure resistor. Therefore, from a very low frequency until 5 kHz, the proposed emulator can accurately mimic the nonlinear property of a real memristor. For any frequency above this range, the emulator acts like a linear resistor. This emulator can be tuned for higher frequencies by changing the time constant of the integrator R_2C that controls the frequency range.

5.2.4   Emulator Implementation for the Generalized Memristor Model (GMM)

To implement an emulator circuit for the GMM model following the design of Fig. 5.8, we need a shaping function circuit, which was not present in the emulator

Fig. 5.10 Experimental results showing the pinched hysteresis loop in the I-V plane for the emulator circuit implemented for the HP memristor model at various frequencies – (**a**) sinusoidal input signal and (**b**) triangular input signal

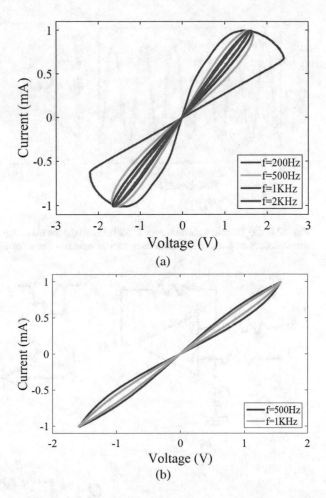

for the HP model. For the GMM model emulator, an exponential function is required. To implement the exponential nonlinearity, we utilized the well-known exponential voltage amplifier circuit (Fig. 5.12b) with a pre-differential to a single-ended converter (Fig. 5.12a). The output voltage of the exponential amplifier is given by Eq. (5.19), where I_{ES} is the reverse saturation current in the order of 10^{13} A, and V_T is the thermal voltage.

$$V_o = I_{ES}R_D\left(1 - e^{\frac{V_{in}}{V_T}}\right) \approx -I_{ES}R_D e^{\frac{V_{in}}{V_T}} \tag{5.19}$$

To realize an asymmetric shaping function, two amplifier circuits with two different transistors having different characteristics were used. One was an NPN, and the other was a PNP transistor, as shown in Fig. 5.12c. The output voltage of the first Op-Amp, V_{o1}, is shown in Eq. (5.20).

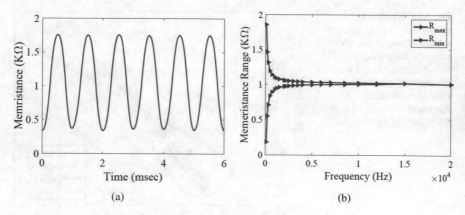

Fig. 5.11 (**a**) Transient memristance at 500 Hz and (**b**) the maximum and the minimum achievable memristances as functions of frequency for the emulator circuit of Fig. 5.9

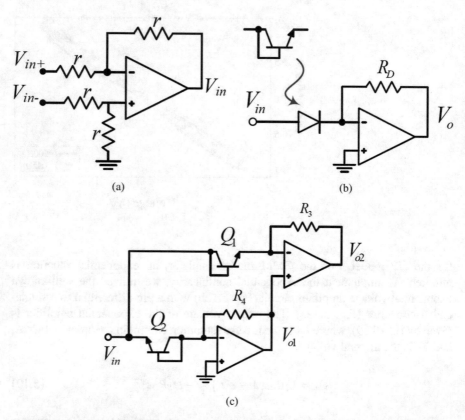

Fig. 5.12 Proposed design of the shaping function circuit for the GMM model – (**a**) a differential to a single-ended converter, (**b**) an exponential amplifier, and (**c**) the schematic diagram of the shaping function circuit

Fig. 5.13 The measured nonlinear curve of the implemented asymmetric shaping function circuit of Fig. 5.12

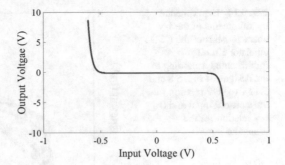

$$V_{o1} \cong I_{ES1} R_4 e^{\frac{-V_{in}}{V_T}} \tag{5.20}$$

The output of the second Op-Amp, V_{o2}, is shown in Eq. (5.21).

$$V_{o2} \cong I_{ES1} R_4 e^{\frac{-V_{in}}{V_T}} - I_{ES2} R_3 e^{\frac{V_{in}}{V_T}} \tag{5.21}$$

where I_{ES3} and I_{ES4} are the saturation current of the transistors connected to R_3 and R_4, respectively. The output voltage of the shaping function is given by Eq. (5.22).

$$V_{o2} \cong -I_{ES1} \left(R_3 e^{\frac{V_{in}}{V_T}} - R_4 e^{\frac{-V_{in}}{V_T}} \right) \tag{5.22}$$

It is worth noting that when $V_{in} > 0$, and $V_{in} < 0$, then $0 < e^{\frac{-V_{in}}{V_T}} < 1$ and $0 < e^{\frac{V_{in}}{V_T}} < 1$, respectively, and gives a relation similar to Eq. (2.11) in Chap. 2, where $A_p = I_{ES} R_3$, $A_n = I_{ES} R_4$, $V_p = \frac{-V_{in}}{V_T}$, and $V_n = \frac{V_{in}}{V_T}$.

Figure 5.13 shows the experimental results of the nonlinear curve of the asymmetric shaping function that is implemented using discrete components. We used PN2222A and PN2906 as diode-connected transistors and TL084 as an amplifier, and the passive elements are $r = 1\ k\Omega$ and $R_3 = R_4 = 1\ k\Omega$ to implement the circuit of Fig. 5.12a. The input voltage is of amplitude 0.6 V at 1 *kHz*, and the DC supply source is ± 9 V. It is evident from Fig. 5.13 that the shape is not symmetric, and its negative part has a higher dead region than the positive part. To realize symmetric shaping function, two transistors with the same characteristics can be used. Moreover, to precisely model, the symmetric shaping function, a MATLAB curve fitting toolbox was used to extract the model parameters. The fitted model is $y = asinh(bx)$, where $a = 5.04 * 10^{-10}$, and $b = -34.7$.

If we insert the shaping function circuit of Fig. 5.12c into the proposed emulator circuit model of Fig. 5.8, we achieve an emulator circuit for the GMM. It is important to note that the exponential amplifier of Fig. 5.12b is inverting. Therefore, a negative integrator is used instead of a positive integrator (shown in Fig. 5.8) to cancel the negative sign and ensure the stability of the feedback. The emulator circuit of Fig. 5.8 for the GMM model was implemented using three AD844s as CCII, one

Fig. 5.14 Experimental measurements of the I-V characteristics of the GMM emulator circuit implemented according to the designs of Fig. 5.8 and 5.12 – (**a**) I-V relation for the sinusoid input and (**b**) I-V relation for the triangular input

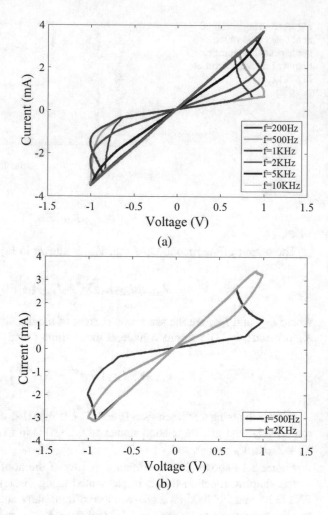

(a)

(b)

AD633 as the multiplier, and passive elements: $R = R_1 = 1K\Omega$, $R_2 = 20K\Omega$, and $C = 5$ nF. Here, the DC supply voltages are ± 9 V. The experimental results for the implemented emulator circuit for the GMM model is shown in Fig. 5.14.

Figure 5.14a reveals that the proposed circuit exhibits the fingerprints of a real memristor for a sinusoidal input at six different frequencies. At low frequencies, the circuit shows nonlinear hysteresis in the (I-V) plane. However, with the increase of frequency of the input signal, this nonlinearity gradually shrinks. Beyond 10 kHz, the emulator starts to behave like a linear resistor, satisfying memristor properties. Figure 5.14b demonstrates the hysteresis loop in the I-V plane of the emulator circuit driven by a triangular input signal at two different frequencies (500 Hz and 2 kHz).

Figure 5.15a shows that with time, the memristance changes for an applied sinusoidal signal with amplitude 2 V and frequency 1 kHz. The memristance spans

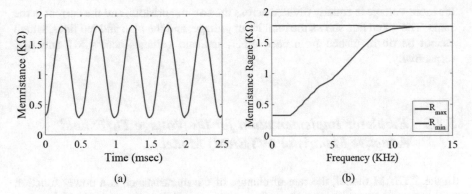

Fig. 5.15 Experimental measurement of memristance of the GMM emulator circuit – (**a**) transient memristance at 1 kHz and (**b**) the maximum and the minimum achievable memristance as a function of frequency

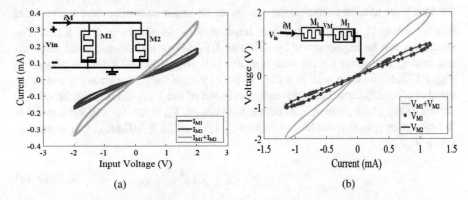

Fig. 5.16 Pinched hysteresis loops in the I-V characteristics of the GMM emulator circuit at 5 kHz – (**a**) two parallel-connected memristors and (**b**) two series-connected memristors

from 0.3 $k\Omega$ up to 1.8 $k\Omega$. Figure 5.15b illustrates that with the increase of frequency, the memristance range decreases until it reaches a critical frequency (10 kHz), where the memristance is constant.

To validate the functionality of the proposed floating emulator circuit, two of the emulators are connected in parallel and in a series to make sure that the current and the voltage are divided equally. Figure 5.16a shows the pinched hysteresis loop in the voltage and current relationship of the proposed emulator circuit for a parallel connection at 1 kHz and 5 kHz. In Fig. 5.16a, the red line represents M1, and the blue line represents M2 behavior with a sinusoidal voltage signal of a 2 Vp-p magnitude. We observed that the current was divided equally between them. The green line represents the total current of the two emulator circuits (M1 + M2) in parallel connection. Figure 5.16b shows the series connection of two memristors, where the input voltage is equally divided across the two memristors, and the current is the same. The green line shows the total input voltage, and the blue and red lines, which cannot be distinguished from each other, indicate voltages across M1 and M2, respectively.

5.2.5   Emulator Implementation for the Voltage ThrEshold Adaptive Memristor (VTEAM) Model

In the VTEAM model, the rate of change of the memristance is a power function with a specific power (α) that depends on the required fitting model [91]. The power relation can be realized by the multiplication operation. From Eq. (2.12) in Chap. 2, we observed that the implemented circuit must have an output voltage: $V_o = k(V_{in}/V_{dc} - 1)^\alpha$. By rearranging this equation, we obtained: $V_o = \left(k_{dc}^\alpha\right) \times (V_{in} - V_{dc})^\alpha$ for which we needed to implement a subtraction and a multiplier circuit. The inner bracket term, $V_x = V_{in} - V_{dc}$, was implemented, as shown in Fig. 5.17a. The subtractor can be implemented by either an OPAMP or a CCII and some resistors. An analog multiplier circuit was implemented using two AD633s, as shown in Fig. 5.17b, to obtain $\alpha = 3$. It is easy to get any required power by using a suitable number of multipliers. The input-output relation of the VTEAM shaping function is shown in Eq. (5.23), where V_o is the output voltage, V_{in} is the input voltage, and β is the multiplier gain that can be demonstrated as $\beta = (R_w + R_Z)/(0.1R_w)$. Consequently, $k = \beta^2 V_{dc}^3$.

$$V_o = \beta^2 (V_{in} - V_{dc})^3 \tag{5.23}$$

Figure 5.17c shows the nonlinear relation between the input and the output of the implemented shaping function circuit. We also observed that the curve has a dead zone, which provided the two states (v_{on} and v_{off}) of the memristor. In the VTEAM model, this dead zone indicates that there will be no change in the memristance if the applied voltage is within this range. This dead region can be controlled in the proposed model by changing the V_{dc}. In our experiments, we used zero V_{dc} to show the effect of power only. By inserting this shaping function into the emulator circuit of Fig. 5.8, the emulator circuit for the VTEAM model can be obtained. For the experimental validation, the proposed emulator circuit for the VTEAM model was implemented by three AD844s as CCII, one AD633 as the voltage multiplier, and some passive elements: $R = R_1 = 1\ k\Omega$, $R_2 = 2\ k\Omega$, and $C = 1\ nF$. The subtractor was implemented by an OPAMP with (TL084, r = 1 $k\Omega$). The DC supply voltages were ± 9 V. The applied voltage signal had an amplitude of 1.5 V.

From the experimental results of Fig. 5.18a, it is evident that the emulator had a pinched hysteresis loop in the I-V plane. The circuit mimics the behavior of the

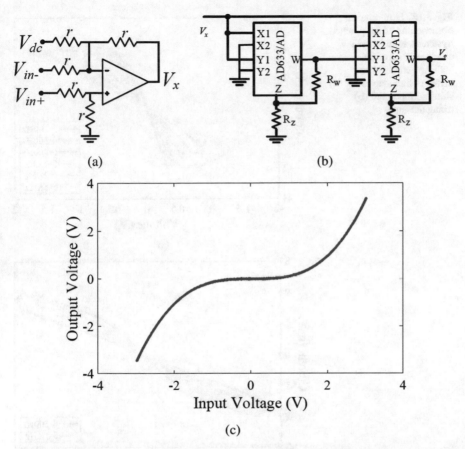

Fig. 5.17 Implementation of the shaping function circuit for the VTEAM model emulator – (**a**) a differential to a single-ended converter, (**b**) the schematic circuit of the shaping function, and (**c**) experimental measurements

VTEAM model. We further observed that with the increase of the applied frequency, the lobe area gradually shrinks. Figure 5.18b shows the nonlinearity of the proposed emulator circuit for a triangular input signal.

Figure 5.19a shows that with time, the memristance varies from 350 Ω to 1.17 kΩ as we apply a sinusoidal signal of 1 kHz frequency and 2 V of amplitude. From Fig. 5.19b, we see that the maximum and the minimum achievable memristances of the emulator change with the applied frequency. We also implemented a series and parallel connections for the emulator for the VTEAM model to study the behavior at different frequencies. Figure 5.20 shows that the results are as expected as in the cases of the GMM emulator circuit's series and parallel connections (see Fig. 5.16).

Fig. 5.18 Experimental observations of the pinched hysteresis loop in the I-V characteristics of the VTEAM emulator at different frequencies for (**a**) sinusoid input and (**b**) triangular input

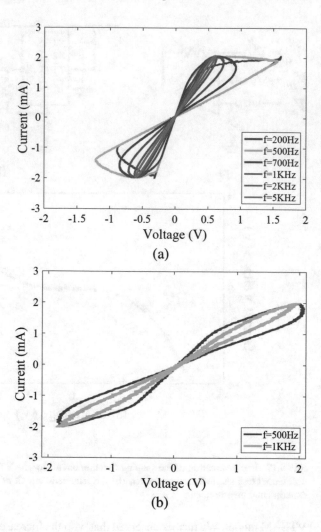

(a)

(b)

5.3   Conclusion

In this chapter, generic and practical emulator circuit's development techniques for the voltage-controlled memristor models are presented. The analytical observations and the experimental results show that the proposed circuits can mimic the nonlinear behavior of real memristors for specific frequency ranges. The proposed circuit model of Fig. 5.8 is a simple memristor emulator that is suitable for use in many digital and analog applications as a two-terminal device. The circuit development technique is generic and practical because it can be configured for most of the

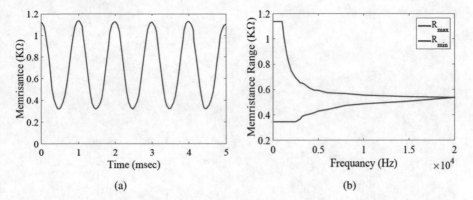

Fig. 5.19 Experimental measurement of memristance of the VTEAM emulator circuit – (**a**) transient memristance and (**b**) the maximum and the minimum achievable memristance as a function of frequency

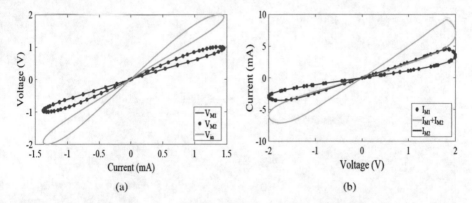

Fig. 5.20 Pinched hysteresis loops in the I-V characteristics of the VTEAM emulator circuit – (**a**) two series-connected memristors at 500 Hz and (**b**) two parallel-connected memristors at 1 kHz

available conceptual memristor models (HP, GMM, and VTEAM), and the frequency ranges can also be changed by using appropriate circuit components. The emulator circuit model of Fig. 5.8 is suitable for implanting both grounded and floating voltage-controlled memristors. These emulator circuits can also be placed both in parallel and series connections. These flexibilities make the proposed emulators very versatile and useful for investigating a wide range of memristor properties, design challenges, and potential applications. Therefore, the proposed emulator circuit's development technique would have a significant impact on the development and educational aspects of this new direction of research.

Chapter 6
Emulator Circuits for the Flux-Controlled Memristive Devices

6.1 Proposed Emulator for a Floating Flux-Controlled Memductor

An emulator circuit for a floating flux-controlled memductor is designed with a voltage difference circuit, a voltage integrator, and an analog multiplier, as shown in Fig. 6.1. Both the voltage difference circuit and the integrator are implemented using the second-generation current conveyor (CCII).

6.1.1 Mathematical Analysis of the Proposed Emulator

The characteristics of an ideal CCII can be represented as in Eq. (6.1).

$$V_Y(t) = V_X(t) \text{ and } i_Z(t) = i_X(t) \tag{6.1}$$

The input current to the circuit can be expressed as in Eq. (6.2).

$$i_{AB}(t) = \frac{V_A - V_B}{R_2} \tag{6.2}$$

Based on Eq. (6.1), the output voltage of the first CCII, V_{ZA}, is as shown in Eq. (6.3). The second CCII works as an integrator, where the current i_{AB} is integrated through the capacitor. Hence, V_{ZB} is given by Eq. (6.4).

$$V_{ZA} = V_{AB} \frac{R_3}{R_2} \tag{6.3}$$

© Springer Nature Switzerland AG 2021
A. G. Alharbi, M. H. Chowdhury, *Memristor Emulator Circuits*,
https://doi.org/10.1007/978-3-030-51882-0_6

Fig. 6.1 Proposed emulator circuit for a floating flux-controlled memductor

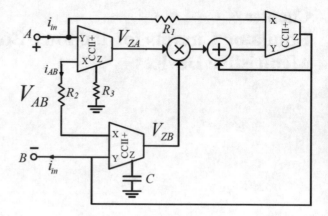

$$V_{ZB} = \frac{-\alpha}{R_2 C} \int_0^t V_{AB}(\tau)\, d\tau + V_Z \tag{6.4}$$

The voltages V_{ZA} and V_{ZB} are multiplied, α is the multiplier constant, and summed to V_Z of the third CCII that represents V_B. Consequently, the voltage of the Y terminal of the third CCII is $V_Y = V_{ZA} + V_{ZB} + V_B$. The input current, I_{in}, is shown in Eq. (6.5).

$$i_{in}(t) = \frac{V_A - (V_{ZA} V_{ZB} + V_B)}{R_1} \tag{6.5}$$

By rearranging this equation, we obtain Eq. (6.6).

$$i_{in}(t) = \frac{V_{AB} - (V_{ZA} V_{ZB})}{R_1} \tag{6.6}$$

This current is mirrored to the input terminal V_B, so the input current to node A is the same as the output current of node B. Hence, this emulator represents a floating memristor emulator. Therefore, the current-voltage relation can be given by Eq. (6.7).

$$i_{in}(t) = V_{AB} \left(\frac{1}{R_1} + \frac{\alpha R_3}{R_2^2 R_1 C} \int_0^t V_{AB}(\tau)\, d\tau \right) \tag{6.7}$$

The input transconductance of the emulator, $G_m = i_{in}(t)/V_{in}(t) = i_{in}/V_{AB}$, represents memductance (memory transconductance), as shown in Eq. (6.8).

$$G_m = \frac{1}{R_1} + \frac{\alpha R_3}{R_2^2 R_1 C} \, \phi\,(t) \tag{6.8}$$

The memductance G_m is a function of the flux ϕ (t). Thus, we refer to this model as the flux-controlled memductance.

6.1.2 Circuit Realization

The proposed design of the emulator for the floating flux-controlled memductor of Fig. 6.1 is implemented in the lab using off-the-shelf components. We used the commercially available AD844 current feedback operational amplifiers (CFOA) as the second-generation current conveyor (CCII) and AD633 as the multiplier. The values of the passive elements are R1 = 10 kΩ, R2 = 22 kΩ, R3 = 10 kΩ, and C = 1 nF, as shown in Fig. 6.2. We used DC supply voltages of ± 9 V. To plot the I-V curves, we sensed the current by using an instrumentation amplifier that measured the differential voltage across the resistance R1. We used the Digilent Electronics Explorer Board to collect the data and PC-based WaveFormsTM software to export the experimental data into MATLAB without alteration to draw the hysteresis loop.

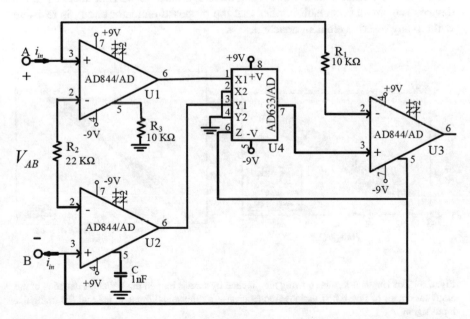

Fig. 6.2 Implementation of the emulator circuit design illustrated in Fig. 6.1 for the floating flux-controlled memductor model

6.1.3   SPICE Simulation and Experimental Validation

Figure 6.3 shows the nonlinearity and the hysteresis loop in the I-V plane of the implemented emulator circuit of Fig. 6.2. At low frequencies, the circuit shows nonlinearity in the (I-V) plane, and at higher frequencies, the nonlinearity gradually disappears. The emulator acts as a nonlinear device and demonstrates a pinched hysteresis loop in the I-V plane for a particular range of frequency as expected. Above a specific frequency, the lobe area of the hysteresis loop tends to shrink. The loop becomes a straight line, and the emulator acts like a linear resistor above a specific frequency, as shown in Fig. 6.3a. The findings indicate that the proposed emulator circuit exhibits all the fingerprints of a memristor listed in [13]. The experimental measurements of Fig. 6.3a are performed for a sinusoidal input signal at different frequencies. Figure 6.3b shows the I-V relations of the proposed emulator circuit for a triangular input signal at 5 *kHz* and 8 *kHz*.

The transient memductance of the emulator circuit of Fig. 6.2 is shown in Fig. 6.4a. It is observed that the memductance is sinusoidal for a sinusoidal input signal. It is well-known that the memristor is resistive and there should not be any phase shift between the current and the voltage through it. From the SPICE simulation, as shown in Fig. 6.4b of the proposed emulator circuit, it is noticed that the current is zero whenever the voltage is zero, which is a signature of the memristor. If there is a phase shift, it would mean that there is a reactive element attached to the device. The simulation result implies that the proposed emulator circuit is resistive without any reactive element attached.

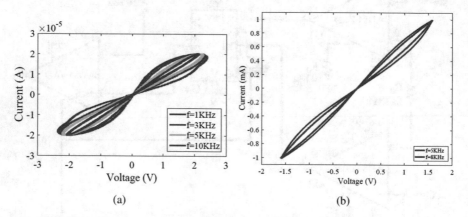

(a)   (b)

Fig. 6.3 Experimental results showing the pinched hysteresis loop in the I-V characteristics of the emulator circuit of Fig. 6.2 at various frequencies – (**a**) sinusoidal input signal and (**b**) triangular input signal

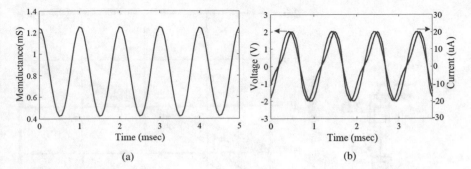

(a) (b)

Fig. 6.4 SPICE transient simulation of the emulator circuit of Fig. 6.2 at 1 *k*Hz – (**a**) transient waveforms of the memductance and (**b**) waveforms of the input voltage $v(t)$ (blue) and the input current $i(t)$ (red)

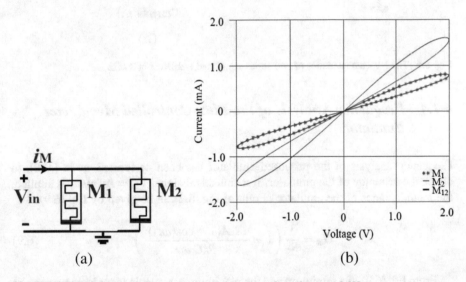

(a) (b)

Fig. 6.5 The I-V characteristics of two parallel-connected emulators at 3 *k*Hz

To demonstrate the functionality and the flexibility of the proposed emulator circuit, two of the emulators are connected in parallel/series to make sure that the current/voltage is divided equally. Figure 6.5 shows the voltage and current relation (pinched hysteresis) in the I-V plane of the proposed emulator circuit for a parallel connection. It is noticed that the current is divided equally between the two emulators. Figure 6.6 shows the series connection of two emulators, where the input voltage is divided equally across the two memristor emulators.

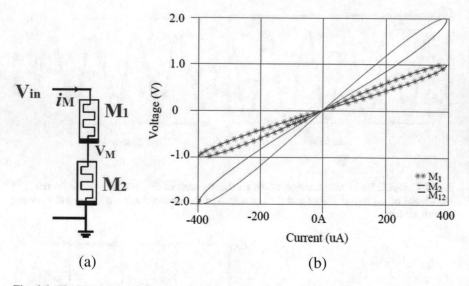

(a)   (b)

Fig. 6.6 The I-V characteristics of two series-connected emulators at 3 *kHz*

6.1.4   Frequency Analysis of the Flux-Controlled Memductor Emulator

Frequency analysis of the proposed emulator has been performed, as in [120], to check the accuracy of the emulator. If a sinusoidal signal $V_{in} = Asin(\omega t)$ is applied, the memductance of the emulator circuit of Fig. 6.2 can be given by Eq. (6.9).

$$G_m = \frac{1}{R_1}\left(1 + \frac{\alpha R_3 A(1 - \cos(\omega t))}{R_2^2 C\,\omega}\right) \tag{6.9}$$

From Eq. (6.9), the minimum and the maximum achievable memductances can be derived as in Eq. (6.10)

$$G_{min} = \frac{1}{R_1} \ and \ G_{max} = \frac{1}{R_1}\left(1 + \frac{2\alpha R_3\,A}{R_2^2 C\,\omega}\right) \tag{6.10}$$

From (6.10), it is evident that with the increase of frequency, ω increases, and the memductance decreases (Fingerprint 2). When ω tends to ∞, Gmax tends to Gmin, which is a constant value leading to non-hysteresis behavior (Fingerprint 3). However, when ω tends to 0, G_{max} tends to ∞, which is not practical since G_{max} saturates to specific values that correspond to R_{on} and R_{off} in the devices.

From Eq. (6.9), it can be observed that the memductance equation has two terms: the first term is a constant and time-invariant resistance, and the second term is a

Fig. 6.7 Frequency-
dependent behavior of the
flux-controlled memductor
emulator of Fig. 6.2 for
$\beta = 0.5$ and $A = 1$

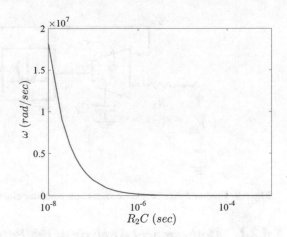

time-varying resistor. The time-varying term is a function of frequency and the time constants of the integrator. The ratio between the magnitude of both terms, β, can be defined as in Eq. (6.11), where $\beta o = \frac{\alpha R_3 A}{\pi R_2}$, $\tau = R_2 C$, and $T = 2\pi/\omega$. The ratio β decreases with the increase of frequency. β tends to zero if the frequency tends to ∞, where the memristor behavior is dominated by the linear term, which is R_1. Also, the hysteresis loop disappears when the time constant (τ) of the integrator is much greater than T.

Figure 6.7 shows the effect of changing the time constants τ with the frequency while maintaining the same ratio $\beta = 0.5$. With the decrease of τ, the higher operating frequency is required to maintain the same value of β_o.

$$\beta = \frac{2\alpha R_3 A}{R_2^2 C \omega} = \beta_0 \frac{T}{\tau} \tag{6.11}$$

6.2   Proposed Emulator for a Floating Flux-Controlled Memristor

In this section, another emulator circuit is presented for a flux-controlled memristor. The proposed emulator circuit is a modification and improved version of the original emulator circuit illustrated in [109]. The proposed emulator circuit, shown in Fig. 6.8, consists of an integrator, a differentiator, and a square function. The integrator and the differentiator are based on two second-generation current conveyors (CCII+s). The square function is used to achieve the required nonlinearity of memristive behavior.

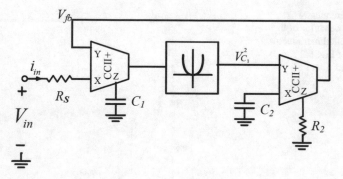

Fig. 6.8 Proposed flux-controlled memristor emulator circuit with a square function

6.2.1   Mathematical Analysis of the Proposed Emulator

The input current (i_{in}) in the circuit of Fig. 6.8 can be found by subtracting the feedback voltage (V_{fb}) from the input voltage (V_{in}), which can be written as in Eq. (6.12).

$$i_{in} = \frac{V_{in} - V_{fb}}{R_s} \tag{6.12}$$

The input current, if imposed in the capacitor C_1, we get the voltage across the capacitor as in Eq. (6.13).

$$V_C = \frac{-1}{R_S C_1} \int_0^t \left(V_{in} - V_{fb} \right) d\tau \tag{6.13}$$

where V_{fb} represents the feedback voltage (output of the second CCII+); this voltage is squared, using the squarer circuit, or multiplied by itself using a multiplier. In our implementation, we used a multiplier. The output voltage of the multiplier is differentiated using the second CCII+ to obtain the feedback voltage, as shown in Eq. (6.14).

$$V_{fb} = \alpha R_2 C_2 \frac{dV_C^2}{dt} \tag{6.14}$$

where α is the multiplier gain. By substituting Eq. (6.14) into Eq. (6.13), the capacitor voltage V_C can be given by Eq. (6.15).

$$V_C = \frac{-1}{R_S C_1} \int_0^t \left(V_{in} - \alpha R_2 C_2 \frac{dV_C^2}{dt} \right) d\tau = \frac{1}{R_S C_1} \left(\alpha R_2 C_2 V_C^2 - \phi_{in} \right) \tag{6.15}$$

By rearranging Eq. (6.15), we get the Eq. (6.16).

$$\alpha R_2 C_2 V_C^2 - R_S C_1 V_C - \phi_{in} = 0 \tag{6.16}$$

It is a second-order equation of V_C, and it can be solved to obtain V_C in terms of the passive elements and known factors as in Eq. (6.17).

$$V_C = \frac{R_S C_1 \pm \sqrt{R_S^2 C_1^2 + 4\alpha R_S C_2 \phi_{in}}}{2\alpha R_2 C_2} \tag{6.17}$$

The feedback voltage is a function of dV_C^2/dt, where the derivative of V_C is given by Eq. (6.18).

$$\frac{dV_C}{dt} = \pm \frac{V_{in}}{\sqrt{R_S^2 C_1^2 + 4\alpha R_2 C_2 \phi_{in}}} \tag{6.18}$$

By applying the chain rule, the derivative of V_C^2 can be found as in Eq. (6.19).

$$\frac{dV_C^2}{dt} = 2V_C \frac{dV_C}{dt} = \pm 2V_C \frac{V_{in}}{\sqrt{R_s^2 C_1^2 + 4\alpha R_2 C_2 \phi_{in}}} \tag{6.19}$$

and the feedback voltage is as shown in Eq. (6.20).

$$V_{fb} = \alpha R_2 C_2 \frac{dV_C^2}{dt} = \left(1 \pm \frac{R_S C_1}{\sqrt{R_s^2 C_1^2 + 4\alpha R_2 C_2 \phi_{in}}}\right) V_{in} \tag{6.20}$$

By substituting the feedback voltage of Eq. (6.20) into Eq. (6.12), and simplifying the expression, the input current can be obtained as in Eq. (6.21) and the input memristance as in Eq. (6.22).

$$i_{in} = \frac{C_1}{\sqrt{R_s^2 C_1^2 + 4\alpha R_2 C_2 \phi_{in}}} \tag{6.21}$$

$$R_m = R_s \sqrt{1 + \frac{4\alpha R_2 C_2 \phi_{in}}{R_s^2 C_1^2}} \tag{6.22}$$

It can be observed that the Eq. (6.22) is the same closed-form solution of the HP model. By comparing this equation and the HP model solution, it can be concluded that the initial memristance is R_s and the memristor speed term, k', can be given by Eq. (6.23).

Fig. 6.9 Effect of changing circuit parameters on the memristor speed for $\alpha = 0.1$

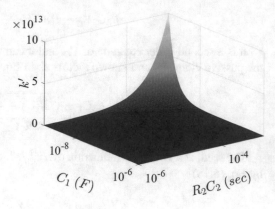

$$k' = \frac{\alpha R_2 C_2}{C_1^2} \tag{6.23}$$

The memristor speed decreases quadratically with the increase of C_1 and linearly increases with the increase of the differentiator time constant, $R_2 C_2$. Figure 6.9 shows a 3D plot of the relation among these three parameters (k', C_1, and $R_2 C_2$).

6.2.2 Circuit Realization

The proposed emulator circuit is simple and implemented using the off-the-shelf components. The commercial version of AD844AN was used as the second-generation current conveyors (CCII), and the square function was implemented using a commercial AD633 (voltage multiplier). The passive element values are $R_s = 1.5\ k\Omega$, $R_2 = 2\ k\Omega$, $C_1 = \mu F$, and $C_2 = 1\ \mu F$ as shown in Figs. 6.10a and 6.11. The DC supply voltages are $\pm 9\ V$. We used the Digilent Electronics Explorer Board and WaveFormsTM software to perform the experimental measurements. The data was exported directly to MATLAB without alterations to draw the hysteresis loop.

6.2.3 SPICE Simulation and Experimental Validation

Figure 6.10b shows the nonlinear behavior of the square function obtained from the SPICE simulation of the implemented square function circuit. If the square function circuit of Fig. 6.10a is plugged into the proposed emulator circuit model of Fig. 6.8, we obtain the implementation of a flux-controlled memristor emulator, as shown in Fig. 6.11. The hysteresis loops derived from the experimental data of the implemented circuit of Fig. 6.11 are shown in Fig. 6.12. It is observed that the emulator has a pinched hysteresis loop in the I-V plane as expected. At low

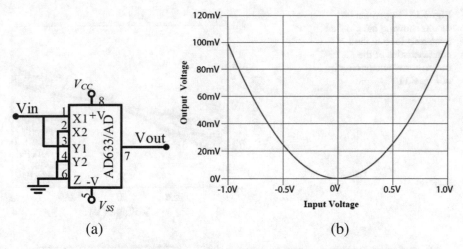

Fig. 6.10 Implementation of the square function: (**a**) circuit diagram and (**b**) SPICE simulation of the nonlinear curve of the square function

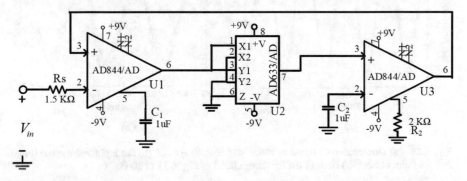

Fig. 6.11 The schematic diagram of the flux-controlled emulator circuit implemented for emulator design illustrated in Fig. 6.8

frequencies, the circuit shows nonlinear hysteresis in the (I-V) plane. However, with the increase of frequency of the input signal, this nonlinearity gradually shrinks. Beyond 2 kHz, the emulator starts to behave like a linear resistor, which satisfies Chua's condition in [5].

By applying a sinusoidal signal of 30 Hz frequency and 1 V amplitude, the change in memeristance in time can be observed as in Fig. 6.13a. It is also noticed that the nature of change in the memristance is sinusoidal as well. Figure 6.13b shows the waveform of the voltage and the current in the emulator circuit of Fig. 6.11. It is observed that there is no phase shift between the current and the voltage. No current exists when the voltage is zero, which validates a significant property of the memristor. Therefore, it can be concluded that the proposed emulator circuit is purely a resistive element.

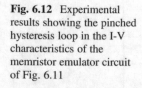

Fig. 6.12 Experimental results showing the pinched hysteresis loop in the I-V characteristics of the memristor emulator circuit of Fig. 6.11

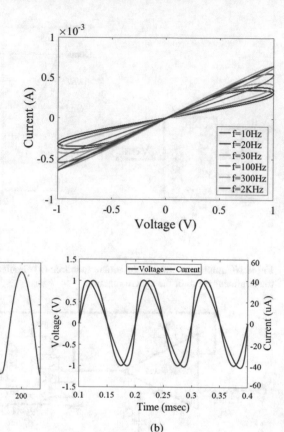

Fig. 6.13 (a) Observation of transient memristance at 30 Hz and (b) the input voltage $v(t)$ (blue) and the input current $i(t)$ (red) of the emulator circuit of Fig. 6.11 at 30 Hz

Figure 6.14 shows the maximum and the minimum achievable memristance, which changes with the frequency. Here, R_{min} is almost constant, and it represents R_S. R_{max} decreases gradually with the applied frequency of the sinusoidal input signal, and at one point, it coincides with R_{min}. This indicates that beyond a specific frequency, the emulator circuit starts to behave like a linear resistor.

6.2.4 Frequency Analysis of the Flux-Controlled Memristor Emulator

As in the previous one, frequency analysis of the proposed flux-controlled memristor emulator is performed to validate the accuracy of the design. By applying a

Fig. 6.14 Measurement of the maximum and minimum achievable memristances as functions of frequency in the implemented emulator circuit of Fig. 6.11

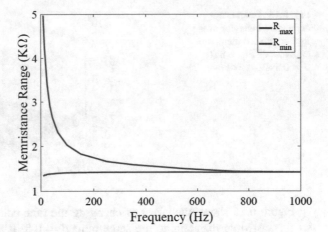

sinusoidal signal, $V_{in} = A\sin(\omega t)$, the memristance can be found as in Eq. (6.24). According to this equation, the minimum and maximum achievable memristance are as shown in Eq. (6.25).

It is evident that with the increase of frequency ω, the memristance decreases (Fingerprint 2). When ω tends to ∞, R_{max} tends to R_{min}, which is a constant value, and it means that there is no hysteresis behavior (Fingerprint 3). However, when ω tends to 0, R_{max} tends to ∞, which is not practical since R_{max} saturates to a specific value, as shown in Fig. 6.14, that corresponds to R_{on} and R_{off} in the device. The memristance equation of Eq. (6.24) has two terms – the first term is a constant resistor, and the second one is a time-varying resistance, which changes with the frequency and time constants of the differentiator and integrator. The ratio between the magnitude of both terms, β, can be defined as in Eq. (6.26).

$$R_m = R_S\sqrt{1 + \frac{4\alpha R_2 C_2 A(1 - \cos{(\omega t)})}{\omega R_s^2 C_1^2}} \tag{6.24}$$

$$R_{min} = R_S \text{ and } R_{max} = R_S\sqrt{1 + \frac{8\alpha R_2 C_2 A}{\omega R_s^2 C_1^2}} \tag{6.25}$$

$$\beta = \frac{8\alpha R_2 C_2}{R_s^2 C_1^2}\frac{A}{\omega} = \beta_0 \frac{\tau_2 T}{\tau_1^2} \tag{6.26}$$

where $\beta o = 8\alpha A/2\pi$, $\tau_1 = RsC1$, $\tau_2 = R_2C_2$, and $T = 2\pi/\omega$.

It is clear that with the increase of frequency, the ratio β would decrease. By examining the expression of β, it can be deduced that β tends to zero if the frequency tends to ∞, where the memristor behavior is dominated by the linear term, which is R_s. The hysteresis loop disappears when the time constant (τ) of the integrator is greater than $\sqrt{T}$. By increasing the differentiator time constant τ, the hysteresis becomes larger.

Fig. 6.15 Frequency-dependent behavior of the flux-controlled memristor emulator of Fig. 6.8 for $\beta = 0.5$ and $A = 1$

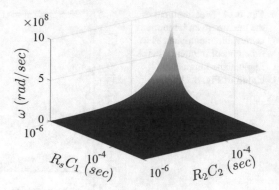

Figure 6.15 shows the effect of changing the time constants τ_1 and τ_2 with the frequency while maintaining the same ratio $\beta = 0.5$. It is observed that with the decrease of τ_1, the higher operating frequency is required to obtain the same value of τ_2 because of the quadratic relation. However, with the decrease of τ_2, a lower operating frequency is required to get the same value of τ_1.

6.3   Conclusion

In this chapter, practical and straightforward memristor emulator circuit development techniques are presented for flux-controlled memductor and memristor. Based on mathematical analysis, SPICE simulation, and experimental measurement, it is observed that the proposed emulator circuits can accurately imitate the behavior of the memristors and satisfy all three fingerprints of a memristor. In the absence of a real solid-state device for memristor, these emulator circuits will be beneficial to investigate the properties and potential applications of the memristor. For example, the proposed emulator circuit model of Fig. 6.1 is a floating memristor emulator, which is suitable for use in many digital and analog applications as a two-terminal device. The proposed circuits are practical and straightforward to implement using widely available and cheaper circuit components compared to many other emulator circuits proposed by different research groups.

Chapter 7
Conclusion and Future Work

Emulator circuits and macro-models are essential techniques for understanding the fundamental properties and potential applications of the memristor. Since no solid-state memristor sample is commercially available in the market, there is a critical need to develop more realistic circuit-based emulators to analyze and investigate the dynamic behavior of this perceived nanoscale element. In this book, we propose generic and practical emulator circuits for the current-controlled and the voltage-controlled memristor models. The proposed emulators, which are designed from the off-the-shelf components, have been implemented and discussed in detail. The proposed emulators better fit the behavior of real memristors compared to the existing emulators. Most of the previous memristor emulators are built on the simple and ideal HP model. Our emulator circuits can be used to emulate the behavior of the electrical nonlinearity of the popular current-controlled memristor models like the Simmons Tunneling Barrier Model (STBM) and the ThrEshold Adaptive Memristor (TEAM) models. Moreover, the proposed emulator circuits have the potential to be compatible with voltage-controlled memristor models such as the Hewlett Packard (HP) Model, the Generalized Memristor Model (GMM), and the Voltage ThrEshold Adaptive Memristor (VTEAM). The proposed emulators are very simple and practical compared to other emulator circuits reported in the literature.

All of the proposed emulators have been designed and implemented using some standard components. The current feedback operational amplifiers (CFOA) as the second-generation current conveyors (CCII) are the most widely used device throughout this book. Other components like an analog multiplier, integrator, passive elements, etc. are also repeated as much as possible in all of the designs. This is done intentionally to keep the design process simple, cost-effective, and easy to implement in the academic labs.

The proposed emulator circuits have been comprehensively studied via mathematical analysis, experimental measurements, and SPICE simulations. The experimental results matched very well with the SPICE simulations, which indicate that the proposed circuits can accurately imitate the behavior of a memristor and satisfy all

© Springer Nature Switzerland AG 2021   89
A. G. Alharbi, M. H. Chowdhury, *Memristor Emulator Circuits*,
https://doi.org/10.1007/978-3-030-51882-0_7

three fingerprints of a memristor. Some quantitative comparisons of the proposed emulators with the most prominent existing emulator circuits are shown.

The proposed emulator models can be configured for both floating and grounded configurations. Therefore, our approach is generic and is more flexible. Researchers are exploring both grounded and floating memristors for diverse applications. For example, many analog applications may require the floating memristor circuit. The previous emulators were designed to fit specific models, which were either voltage-controlled or current-controlled models. Since the memristor's state is stored as a charge in a capacitor, any memristor emulator should be considered as nonvolatile ideally. These emulators were built with an analog integrator where the integrator output represents the state variable that is stored in a capacitor. However, practically, the voltage across the capacitor should be continuously refreshed because the capacitor has a leaking resistance that discharges the state voltage. Therefore, the voltage-controlled emulators are suitable only for continuous analog applications, and it is not wise to use those for digital circuits. The capacitors also need to have a very high-quality factor. Our proposed floating emulators can be used in different circuits in series, parallel, and hybrid configurations without any issue.

Since most of the proposed emulators are being investigated for lower frequency ranges, the parasitic elements have negligible effects. Therefore, we neglected parasitic effects in our analysis. However, if we include these parasitic elements, the maximum working speed of the memristor will be limited. Future work will focus on developing memristor emulators for high-frequency ranges. The reliability and scalability of memristor emulator circuits for different frequency ranges could also be a new research direction.

The actual effectiveness of any model can only be justified if the margin of error between data generated by the proposed model and real data obtained from a fabricated memristive device or circuit is low. However, for memristors, this type of validation is not yet possible because there is no actual solid-state or other types of memristive devices (like the transistor, resistor, inductor, or capacitor) in the real electronics world. In the last decade, researchers have been utilizing the HP model to understand memristor behavior. However, the implemented HP memristive device and relevant models failed to mimic the nonlinear behavior of the memristor. As a consequence, none of the existing work on memristor models and emulators could provide any comparison with reliable fabricated data. The research community is still trying to figure out what would be the feasible materials and physical and electrical characteristics of the memristors. While waiting for a real memristive device to appear, researchers are trying to come up with better macro-models or emulator circuits to analyze and understand the properties and behaviors of memristors for different application environments.

Since we do not have any real memristive device data in hand, we have generated data from the proposed mathematical model and relevant SPICE simulation. We compared these data with the results obtained from the proposed emulator circuit implemented on a PCB board using discrete circuit components. At this stage, our focus is to validate the models and the emulator circuits. In the future, when the research community figures out how to implement and integrate actual memristive devices inside the integrated circuit chips, we can focus on developing different configurations of the device based on our or other models.

We can conclude that in the absence of a real memristive device, the proposed emulator circuit development techniques of this book can be used to investigate different analog and digital applications of the memristor. The proposed emulator circuits are simple and straightforward to implement in the lab for educational purposes.

Bibliography

1. Alexander, C., & Sadiku, M. (2008). *Fundamentals of electric circuits* (4th ed., pp. 1–30). New York: McGraw-Hill Higher Education.
2. Chua, L. O. (1971). Memristor the missing circuit element. *IEEE Transactions on Circuit Theory, 18*(5), 507–519.
3. Strukov, D. B., Snider, G. S., Stewart, D. R., & Williams, R. S. (2008). The missing memristor found. *Nature, 453*(7191), 80–83.
4. Memristor and Memristive Systems Symposium. (2008). University of California. [Online]. https://www.youtube.com/watch?v=QFdDPzcZwbs. Accessed 26 Oct 2017.
5. Chua, L. O. (2014). If its pinched it's a memristor. In *Memristors and memristive systems* (pp. 17–90). New York: Springer.
6. Widrow, B., et al. (1960). *Adaptive "adaline" neuron using Chemical "memistors"*. Stanford University, Stanford Electronics Laboratories. Technical report, 1553-2.
7. Adhikari, S. P., & Kim, H. (2014). Why are memristor and memistor different devices? In *Memristor networks* (pp. 95–112). Cham: Springer.
8. Prodromakis, T., Toumazou, C., & Chua, L. O. (2012). Two centuries of memristors. *Nature Materials, 11*(6), 478–481.
9. Zidan, M. A. (2015). *Memristor circuits and systems*. Ph.D. thesis.
10. Chua, L. O. (2011). Resistance switching memories are memristors. *Applied Physics A, 102* (4), 765–783.
11. Chua, L. O., & Kang, S. M. (1976). Memristive devices and systems. *Proceedings of the IEEE, 64*(2), 209–223.
12. Waser, R., Dittmann, R., Staikov, G., & Szot, K. (2009). Redox-based resistive switching memories–nanoionic mechanisms, prospects, and challenges. *Advanced Materials, 21*(25–26), 2632–2663.
13. Adhikari, S. P., Sah, M. P., Kim, H., & Chua, L. O. (2013). Three fingerprints of memristor. *IEEE Transactions on Circuits and Systems I, 60*(11), 3008–3021.
14. Biolek, D., Biolek, Z., Biolkova, V., & Kolka, Z. (2013). *Some fingerprints of ideal memristors*. In IEEE International Symposium on Circuits and Systems (ISCAS), pp. 201–204.
15. Williams, R. S. (2008). How we found the missing memristor. *IEEE Spectrum, 45*(12), 28–35.
16. Argall, F. (1968). Switching phenomena in titanium oxide thin films. *Solid-State Electronics, 11*(5), 535–541.
17. Chabi, D., Wang, Z., Bennett, C., Klein, J.-O., & Zhao, W. (2015). Ultrahigh density memristor neural crossbar for on-chip supervised learning. *IEEE Transactions on Nanotechnology, 14*(6), 954–962.

© Springer Nature Switzerland AG 2021
A. G. Alharbi, M. H. Chowdhury, *Memristor Emulator Circuits*,
https://doi.org/10.1007/978-3-030-51882-0

18. Kim, K.-H., Gaba, S., Wheeler, D., Cruz-Albrecht, J. M., Hussain, T., Srinivasa, N., & Lu, W. (2011). A functional hybrid memristor crossbar-array/CMOS system for data storage and neuromorphic applications. *Nano Letters, 12*(1), 389–395.

19. Qureshi, M., Yi, W., Medeiros-Ribeiro, G., & Williams, R. (2012). AC sense technique for memristor crossbar. *Electronics Letters, 48*(13), 757–758.

20. Qureshi, M. S., Pickett, M., Miao, F., & Strachan, J. P. (2011). *CMOS interface circuits for reading and writing memristor crossbar array.* In IEEE international symposium on Circuits and systems (ISCAS), pp. 2954–2957.

21. Zidan, M. A., Omran, H., Sultan, A., Fahmy, H. A., & Salama, K. N. (2015). Compensated readout for high-density MOS-gated memristor crossbar array. *IEEE Transactions on Nanotechnology, 14*(1), 3–6.

22. Adam, G. C., Hoskins, B. D., Prezioso, M., Merrikh-Bayat, F., Chakrabarti, B., & Strukov, D. B. (2017). 3-D memristor crossbars for analog and neuromorphic computing applications. *IEEE Transactions on Electron Devices, 64*(1), 312–318.

23. Hu, M., Li, H., Chen, Y., Wu, Q., Rose, G. S., & Linderman, R. W. (2014). Memristor crossbar-based neuromorphic computing system: A case study. *IEEE Transactions on Neural Networks and Learning Systems, 25*(10), 1864–1878.

24. Kim, Y., Zhang, Y., & Li, P. (2012). *A digital neuromorphic VLSI architecture with memristor crossbar synaptic array for machine learning*, In IEEE International SOC Conference (SOCC), pp. 328–333.

25. Olshausen, B. A., & Rozell, C. J. (2017). Neuromorphic computation: Sparse codes from memristor grids. *Nature Nanotechnology, 12*, 722–723.

26. Prezioso, M., Merrikh-Bayat, F., Hoskins, B., Adam, G., Likharev, K. K., & Strukov, D. B. (2015). Training and operation of an integrated neuromorphic network based on metal-oxide memristors. *Nature, 521*, 61–64.

27. Truong, S. N., Van Pham, K., Yang, W., & Min, K.-S. (2016). Sequential Memristor crossbar for neuromorphic pattern recognition. *IEEE Transactions on Nanotechnology, 15*(6), 922–930.

28. Zidan, M., Jeong, Y., Shin, J. H., Du, C., Zhang, Z., & Lu, W. (2018). Field programmable crossbar array (FPCA) for reconfigurable computing. *IEEE Transactions on Multi-Scale Computing Systems, 4*(4), 698–710.

29. Adhikari, S. P., Yang, C., Kim, H., & Chua, L. O. (2012). Memristor bridge synapse based neural network and its learning. *IEEE Transactions on Neural Networks and Learning Systems, 23*(9), 1426–1435.

30. Bilotta, E., Pantano, P., & Vena, S. (2017). Speeding up cellular neural network processing ability by embodying memristors. *IEEE Transactions on Neural Networks and Learning Systems, 28*(5), 1228–1232.

31. Kim, H., Sah, M. P., Yang, C., Roska, T., & Chua, L. O. (2012). Neural synaptic weighting with a pulse-based memristor circuit. *IEEE Transactions on Circuits and Systems I: Regular Papers, 59*(1), 148–158.

32. Li, T., Duan, S., Liu, J., Wang, L., & Huang, T. (2016). A spintronic memristor-based neural network with radial basis function for robotic manipulator control implementation. *IEEE Transactions on Systems, Man, and Cybernetics: Systems, 46*(4), 582–588.

33. Wu, A., & Zeng, Z. (2012). Dynamic behaviors of memristor-based recurrent neural networks with time-varying delays. *Neural Networks, 36*, 1–10.

34. Corinto, F., Ascoli, A., & Gilli, M. (2011). Nonlinear dynamics of memristor oscillators. *IEEE Transactions on Circuits and Systems I: Regular Papers, 58*(6), 1323–1336.

35. Corinto, F., & Forti, M. (2017). *Nonlinear dynamics of memristor oscillators via the flux charge analysis method.* IEEE International Symposium on Circuits and Systems (ISCAS), pp. 1–4.

36. Takahashi, Y., Sekine, T., & Yokoyama, M. (2017). Memristor-based pseudo-random pattern generator using relaxation oscillator. *IEEJ Transactions on Electrical and Electronic Engineering, 12*(6), 963–964.

37. Talukdar, A., Radwan, A. G., & Salama, K. N. (2011). Generalized model for memristor based Wien family oscillators. *Microelectronics Journal, 42*(9), 1032–1038.
38. Yu, D., Zhou, Z., Iu, H. H.-C., Fernando, T., & Hu, Y. (2016). A coupled Memcapacitor emulator-based relaxation oscillator. *IEEE Transactions on Circuits and Systems II: Express Briefs, 63*(12), 1101–1105.
39. Zidan, M. A., Omran, H., Radwan, A. G., & Salama, K. N. (2011). Memristor-based reactance-less oscillator. *Electronics Letters, 47*(22), 1220–1221.
40. Iu, H. H.-C., Yu, D., Fitch, A. L., Sreeram, V., & Chen, H. (2011). Controlling chaos in a memristor based circuit using a twin-T notch filter. *IEEE Transactions on Circuits and Systems I: Regular Papers, 58*(6), 1337–1344.
41. Pershin, Y. V., & Di Ventra, M. (2010). Practical approach to programmable analog circuits with memristors. *IEEE Transactions on Circuits and Systems I: Regular Papers, 57*(8), 1857–1864.
42. Shin, S., Kim, K., & Kang, S.-M., (2009). *Memristor-based fine resolution programmable resistance and its applications*. In IEEE International Conference on Communications, Circuits and Systems (ICCCAS), pp. 948–951.
43. Shin, S., Kim, K., & Kang, S.-M. (2011). Memristor applications for programmable analog ICs. *IEEE Transactions on Nanotechnology, 10*(2), 266–274.
44. Wang, X., Iu, H. H., Wang, G., & Liu, W. (2016). Study on time domain characteristics of memristive RLC series circuits. *Circuits, Systems, and Signal Processing, 35*(11), 4129–4138.
45. Zha, J., Huang, H., Huang, T., Cao, J., Alsaedi, A., & Alsaadi, F. E. (2017). A general memristor model and its applications in programmable analog circuits. *Neurocomputing, 267*, 134–140.
46. Zha, J., Huang, H., & Liu, Y. (2016). A novel window function for memristor model with application in programming analog circuits. *IEEE Transactions on Circuits and Systems II: Express Briefs, 63*(5), 423–427.
47. Borghetti, J., Li, Z., Straznicky, J., Li, X., Ohlberg, D. A., Wu, W., Stewart, D. R., & Williams, R. S. (2009). A hybrid nanomemristor/transistor logic circuit capable of self-programming. *Proceedings of the National Academy of Sciences, 106*(6), 1699–1703.
48. Gao, L., Alibart, F., & Strukov, D. B. (2013). Programmable CMOS/memristor threshold logic. *IEEE Transactions on Nanotechnology, 12*(2), 115–119.
49. Guckert, L., & Swartzlander, E. E. (2017). MAD gates-memristor logic design using driver circuitry. *IEEE Transactions on Circuits and Systems II: Express Briefs, 64*(2), 171–175.
50. Guckert, L., & Swartzlander, E. E. (2017). Optimized memristor-based multipliers. *IEEE Transactions on Circuits and Systems I: Regular Papers, 64*(2), 373–385.
51. Owlia, H., Keshavarzi, P., & Rezai, A. (2014). A novel digital logic implementation approach on nanocrossbar arrays using memristor-based multiplexers. *Microelectronics Journal, 45*(6), 597–603.
52. Papandroulidakis, G., Vourkas, I., Abusleme, A., Sirakoulis, G. C., & Rubio, A. (2017). Crossbar-based Memristive logic-in-memory architecture. *IEEE Transactions on Nanotechnology, 16*(3), 491–501.
53. Pershin, Y. V., Shevchenko, S. N., & Nori, F. (2016). Memristive Sisyphus circuit for clock signal generation. *Scientific Reports, 6*, 26155.
54. Vourkas, I., & Sirakoulis, G. C. (2016). Emerging memristor-based logic circuit design approaches: A review. *IEEE Circuits and Systems Magazine, 16*(3), 15–30.
55. Xia, Q., Robinett, W., Cumbie, M. W., Banerjee, N., Cardinali, T. J., Yang, J. J., Wu, W., Li, X., Tong, W. M., Strukov, D. B., et al. (2009). Memristor- CMOS hybrid integrated circuits for reconfigurable logic. *Nano Letters, 9*(10), 3640–3645.
56. Zheng, J., Zeng, Z., & Zhu, Y. (2017) *Memristor-based nonvolatile synchronous flip-flop circuits*. In International Conference on Information Science and Technology (ICIST), IEEE, pp. 504–508.
57. Ascoli, A., Tetzlaff, R., Corinto, F., Mirchev, M., & Gilli, M., (2013) *Memristor-based filtering applications*. In 14th Latin American Test Workshop (LATW), IEEE, pp. 1–6.

58. Chew, Z., & Li, L. (2012). Printed circuit board based memristor in adaptive lowpass filter. *Electronics Letters, 48*(25), 1610–1611.
59. Driscoll, T., Quinn, J., Klein, S., Kim, H.-T., Kim, B., Pershin, Y. V., Di Ventra, M., & Basov, D. (2010). Memristive adaptive filters. *Applied Physics Letters, 97*(9), 093502.
60. Merrikh-Bayat, F., & Bagheri-Shouraki, S. (2011). Mixed analog-digital crossbar-based hardware implementation of sign–sign LMS adaptive filter. *Analog Integrated Circuits and Signal Processing, 66*(1), 41–48.
61. Volos, C., Vaidyanathan, S., Pham, V.-T., Nistazakis, H., Stouboulos, I., Kyprianidis, I., & Tombras, G. (2017). Adaptive control and synchronization of a Memristor based Shinrikis system. In *Advances in Memristors, Memristive devices and systems* (pp. 237–261). Cham: Springer.
62. Bao, B., Liu, Z., & Xu, J. (2010). Steady periodic memristor oscillator with transient chaotic behaviours. *Electronics Letters, 46*(3), 237–238.
63. Biolek, Z., Biolek, D., & Biolkova, V. (2009). SPICE model of Memristor with nonlinear dopant drift. *Radioengineering, 18*(2), 1087.
64. Driscoll, T., Pershin, Y., Basov, D., & Di Ventra, M. (2011). Chaotic memristor. *Applied Physics A: Materials Science & Processing, 102*(4), 885–889.
65. Kumar, S., Strachan, J. P., & Williams, R. S. (2017). Chaotic dynamics in nanoscale NbO2 Mott memristors for analogue computing. *Nature, 548*(7667), 318–321.
66. Zheng, C.-D., & Xian, Y. (2016). On synchronization for chaotic memristor-based neural networks with time-varying delays. *Neurocomputing, 216*, 570–586.
67. Abuelmaatti, M. T., & Khalifa, Z. J. (2014). A new memristor emulator and its application in digital modulation. *Analog Integrated Circuits and Signal Processing, 80*(3), 577–584.
68. Biolek, D., Biolkova, V., & Kolka, Z. (2014). *Memristive systems for analog signal processing*. In IEEE International Symposium on Circuits and Systems (ISCAS), pp. 2588–2591.
69. Elashkar, N., Aboudina, M., Fahmy, H. A., Ibrahim, G. H., & Khalil, A. H. (2016). Memristor based BPSK and QPSK demodulators with nonlinear dopant drift model. *Microelectronics Journal, 56*, 17–24.
70. Elashkar, N., Ibrahim, G., Aboudina, M., Fahmy, H., & Khalil, A. (2016). *All-passive memristor-based 8-QAM and BFSK demodulators using linear dopant drift model*. In 5th International Conference on Electronic Devices, Systems and Applications (ICEDSA), IEEE, pp. 1–4.
71. Goknar, I. C., Oncul, F., & Minayi, E. (2013). New memristor applications: AM, ASK, FSK, and BPSK modulators. *IEEE Antennas and Propagation Magazine, 55*(2), 304–313.
72. Sanchez-Lopez, C., Aguila-Cuapio, L., Carro-Perez, I., & Gonzalez-Hernandez, H. (2016). *High-level simulation of an FSK modulator based on memconductor*. In Argentine Conference of MicroNanoelectronics, Technology and Applications (CAMTA), IEEE, pp. 1–5.
73. Vavra, J., & Biolek, D. (2017). An envelope detector based on Memristive systems. *Journal of Telecommunication, Electronic and Computer Engineering (JTEC), 9*(2–7), 183–186.
74. Chanthbouala, A., Garcia, V., Cherifi, R. O., Bouzehouane, K., Fusil, S., Moya, X., Xavier, S., Yamada, H., Deranlot, C., Mathur, N. D., et al. (2012). A ferroelectric memristor. *Nature Materials, 11*(10), 860–864.
75. Mehonic, A., Cueff, S., Wojdak, M., Hudziak, S., Jambois, O., Labbe, C., Garrido, B., Rizk, R., & Kenyon, A. J. (2012). Resistive switching in silicon suboxide films. *Journal of Applied Physics, 111*(7), 074507.
76. Yang, Y., Choi, S., & Lu, W. (2013). Oxide heterostructure resistive memory. *Nano Letters, 13*(6), 2908–2915.
77. Chen, Y., Liu, G., Wang, C., Zhang, W., Li, R.-W., & Wang, L. (2014). Polymer memristor for information storage and neuromorphic applications. *Materials Horizons, 1*(5), 489–506.
78. Erokhin, V., & Fontana, M. P. (2008). Electrochemically controlled polymeric device: A memristor (and more) found two years ago. *arXiv preprint, arXiv*, 0807.0333.

79. Yilmaz, Y., & Mazumder, P. (2012). *Programmable quantum-dots memristor based architecture for image processing*. In 12th IEEE Conference on Nanotechnology (IEEE-NANO), pp. 1–4.
80. Wang, X., Chen, Y., Xi, H., Li, H., & Dimitrov, D. (2009). Spintronic memristor through spin-torque-induced magnetization motion. *IEEE Electron Device Letters, 30*(3), 294–297.
81. Sangwan, V. K., Jariwala, D., Kim, I. S., Chen, K.-S., Marks, T. J., Lauhon, L. J., & Hersam, M. C. (2015). Gate-tunable memristive phenomena mediated by grain boundaries in single-layer MoS2. *Nature Nanotechnology, 10*(5), 403–406.
82. Porro, S., Accornero, E., Pirri, C. F., & Ricciardi, C. (2015). Memristive devices based on graphene oxide. *Carbon, 85*, 383–396.
83. Russo, P., Xiao, M., & Zhou, N. Y. (2017). Carbon nanowalls: A new material for resistive switching memory devices. *Carbon, 120*, 54–62.
84. Abunahla, H., & Mohammad, B. (2018). Memristor device overview. In *Memristor technology: Synthesis and modeling for sensing and security applications* (pp. 1–29). Cham: Springer.
85. Mohanty, S. P. (2013). Memristor: From basics to deployment. *IEEE Potentials, 32*(3), 34–39.
86. Wang, L., Yang, C., Wen, J., Gai, S., & Peng, Y. (2015). Overview of emerging memristor families from resistive memristor to spintronic memristor. *Journal of Materials Science: Materials in Electronics, 26*(7), 4618–4628.
87. Pickett, M. D., Strukov, D. B., Borghetti, J. L., Yang, J. J., Snider, G. S., Stewart, D. R., & Williams, R. S. (2009). Switching dynamics in titanium dioxide memristive devices. *Journal of Applied Physics, 106*(7), 074508.
88. Kvatinsky, S., Friedman, E. G., Kolodny, A., & Weiser, U. C. (2013). TEAM: ThrEshold adaptive memristor model. *IEEE Transactions on Circuits and Systems I: Regular Papers, 60* (1), 211–221.
89. Yakopcic, C., Taha, T. M., Subramanyam, G., & Pino, R. E. (2012). Memristor SPICE modeling. In *Advances in neuromorphic Memristor science and applications* (pp. 211–244). New York: Springer.
90. Yakopcic, C., Taha, T. M., Subramanyam, G., Pino, R. E., & Rogers, S. (2011). A memristor device model. *IEEE Electron Device Letters, 32*(10), 1436–1438.
91. Kvatinsky, S., Ramadan, M., Friedman, E. G., & Kolodny, A. (2015). VTEAM: A general model for voltage-controlled memristors. *IEEE Transactions on Circuits and Systems II: Express Briefs, 62*(8), 786–790.
92. Joglekar, Y. N., & Wolf, S. J. (2009). The elusive memristor: Properties of basic electrical circuits. *European Journal of Physics, 30*(4), 661.
93. Biolek, Z., Biolek, D., & Biolkova, V. (2009). SPICE model of Memristor with nonlinear dopant drift. *Radioengineering, 18*(2), 1087.
94. Prodromakis, T., Peh, B. P., Papavassiliou, C., & Toumazou, C. (2011). A versatile memristor model with nonlinear dopant kinetics. *IEEE Transactions on Electron Devices, 58*(9), 3099–3105.
95. Elgabra, H., Farhat, I. A., Al Hosani, A. S., Homouz, D., & Mohammad, B., (2012). *Mathematical modeling of a memristor device*. In International IEEE Conference on Innovations in Information Technology (IIT), pp. 156–161.
96. Yang, J. J., Pickett, M. D., Li, X., Ohlberg, D. A., Stewart, D. R., & Williams, R. S. (2008). Memristive switching mechanism for metal/oxide/metal nanodevices. *Nature Nanotechnology, 3*(7), 429–433.
97. Abdalla, H., & Pickett, M. D. (2011). *SPICE modeling of memristors*. In IEEE International Symposium on Circuits and Systems (ISCAS), pp. 1832–1835.
98. Biolek, D., Biolkova, V., & Kolka, Z. (2017). *Modified MIM model of titanium dioxide memristor for reliable simulations in SPICE*. In IEEE, 14th International Conference on Synthesis, Modeling, Analysis and Simulation Methods and Applications to Circuit Design (SMACD), pp. 1–4.

99. Pershin, Y. V., & Di Ventra, M. (2012). SPICE model of memristive devices with threshold. *arXiv preprint, arXiv*, 1204.2600.
100. Vourkas, I., Batsos, A., & Sirakoulis, G. C. (2015). SPICE modeling of nonlinear memristive behavior. *International Journal of Circuit Theory and Applications, 43*(5), 553–565.
101. Yakopcic, C., Taha, T. M., Subramanyam, G., & Pino, R. E. (2013). Generalized memristive device SPICE model and its application in circuit design. *IEEE Transactions on Computer-Aided Design of Integrated Circuits and Systems, 32*(8), 1201–1214.
102. Zhang, Y., Zhang, X., & Yu, J. (2009). *Approximated SPICE model for memristor*. In IEEE International Conference on Communications, Circuits and Systems (ICCCAS), pp. 928–931.
103. Patterson, G., Sune, J., & Miranda, E. (2017). Voltage-driven hysteresis model for resistive switching: SPICE modeling and circuit applications. *IEEE Transactions on Computer-Aided Design of Integrated Circuits and Systems, 36*(12), 2044–2051.
104. Garcia-Redondo, F., Gowers, R. P., Crespo-Yepes, A., Lopez-Vallejo, M., & Jiang, L. (2016). SPICE compact modeling of bipolar/unipolar memristor switching governed by electrical thresholds. *IEEE Transactions on Circuits and Systems I: Regular Papers, 63*(8), 1255–1264.
105. Biolek, D. (2014). Memristor emulators. In *Memristor networks* (pp. 487–503). Cham: Springer.
106. Kolka, Z., Biolek, D., & Biolkova, V. (2012). Hybrid modelling and emulation of mem-systems. *International Journal of Numerical Modelling: Electronic Networks, Devices and Fields, 25*(3), 216–225.
107. Kolka, Z., Biolkova, V., & Biolek, D. (2014). *On hybrid emulation of mem-systems*. In IEEE Proceedings of the 2014 European modelling symposium, Computer Society, pp. 490–494.
108. Biolek, D., Bajer, J., Biolkova, V., & Kolka, Z. (2011). *Mutators for transforming nonlinear resistor into memristor*. In 20th European Conference on Circuit Theory and Design (ECCTD), IEEE, pp. 488–491.
109. Abuelmaatti, M. T., & Khalifa, Z. J. (2015). A continuous-level memristor emulator and its application in a multivibrator circuit. *AEU-International Journal of Electronics and Communications, 69*(4), 771–775.
110. Kim, H., Sah, M. P., Yang, C., Cho, S., & Chua, L. O. (2012). Memristor emulator for memristor circuit applications. *IEEE Transactions on Circuits and Systems I: Regular Papers, 59*(10), 2422–2431.
111. Asapu, S., & Pershin, Y. V. (2015). Electromechanical emulator of memristive systems and devices. *IEEE Transactions on Electron Devices, 62*(11), 3678–3684.
112. Liu, W., Wang, F.-Q., & Ma, X.-K. (2015). A unified cubic flux-controlled memristor: Theoretical analysis, simulation and circuit experiment. *International Journal of Numerical Modelling: Electronic Networks, Devices and Fields, 28*(3), 335–345.
113. Muthuswamy, B., & Kokate, P. P. (2009). Memristor-based chaotic circuits. *IETE Technical Review, 26*(6), 417–429.
114. Nguyen, V. H., Sohn, K. Y., & Song, H. (2016). On-printed circuit board emulator with controllability of pinched hysteresis loop for nanoscale TiO 2 thin-film memristor device. *Journal of Computational Electronics, 15*(3), 993–1002.
115. Zhong, G.-Q. (1994). Implementation of Chua's circuit with a cubic nonlinearity. *IEEE Transactions on Circuits and Systems-Part I-Fundamental Theory and Applications, 41*(12), 934–940.
116. Elwakil, A. S., Fouda, M. E., & Radwan, A. G. (2013). A simple model of double-loop hysteresis behavior in memristive elements. *IEEE Transactions on Circuits and Systems II: Express Briefs, 60*(8), 487–491.
117. Fitch, A. L., Iu, H. H.-C., Wang, X., Sreeram, V., & Qi, W. (2012). *Realization of an analog model of memristor based on light dependent resistor*. In IEEE International Symposium on Circuits and Systems (ISCAS), pp. 1139–1142.
118. Kumngern, M., & Moungnoul, P. (2015). *A memristor emulator circuit based on operational transconductance amplifiers*. In IEEE 12th International Conference on Electrical Engineering/Electronics, Computer, Telecommunications and Information Technology (ECTI-CON), pp. 1–5.

119. Alharbi, A. G., Fouda, M. E., & Chowdhury, M. H. (2015). *A novel memristor emulator based only on an exponential amplifier and ccii+*. In IEEE International Conference on Electronics, Circuits, and Systems (ICECS), pp. 376–379.

120. Sanchez-Lopez, C., Mendoza-Lopez, J., Carrasco-Aguilar, M., & Muniz-Montero, C. (2014). A floating analog memristor emulator circuit. *IEEE Transactions on Circuits and Systems II: Express Briefs, 61*(5), 309–313.

121. Minayi, E., & Goknar, I. C. (2013). *Realization of a 4-port generalized mutator and its application to memstor1 simulations*. In 8th International Conference on Electrical and Electronics Engineering (ELECO), IEEE, pp. 5–8.

122. Yesil, A., Babacan, Y., & Kacar, F. (2014). A new DDCC based memristor emulator circuit and its applications. *Microelectronics Journal, 45*(3), 282–287.

123. Koymen, I., & Drakakis, E. M. (2014). *CMOS-based nanopower memristor dynamics emulator*. In 14th International Workshop on Cellular Nanoscale Networks and their Applications (CNNA), IEEE, pp. 1–2.

124. Kuntman, H., et al. (2012). *A new CMOS based memristor implementation*. In IEEE, International Conference on Applied Electronics (AE), pp. 345–348.

125. Yener, S. C., & Kuntman, H. H. (2014). Fully CMOS memristor based chaotic circuit. *Radioengineering, 23*(4), 1140–1149.

126. Alharbi, A. G., Khalifa, Z. J., Fouda, M. E., & Chowdhury, M. H. (2015). *A new simple emulator circuit for current controlled memristor*. In IEEE International Conference on Electronics, Circuits, and Systems (ICECS), pp. 288–291.

127. Alharbi, A. G., Fouda, M. E., & Chowdhury, M. H. (2015). *Memristor emulator based on practical current controlled mode*. In 58th International Midwest Symposium on Circuits and Systems (MWSCAS), IEEE, pp. 1–4.

128. Yu, D., Iu, H. H.-C., Fitch, A. L., & Liang, Y. (2014). A floating memristor emulator based relaxation oscillator. *IEEE Transactions on Circuits and Systems I: Regular Papers, 61*(10), 2888–2896.

129. Alharbi, A. G., Fouda, M. E., & Chowdhury, M. H. (2017). A novel flux-controlled Memristive emulator for analog applications. In *Advances in Memristors, Memristive devices and systems* (pp. 493–511). Cham: Springer.

130. Abuelmaatti, M. T., & Khalifa, Z. J. (2016). A new floating memristor emulator and its application in frequency-to-voltage conversion. *Analog Integrated Circuits and Signal Processing, 86*(1), 141–147.

131. Budhathoki, R. K., Sah, M. P., Adhikari, S. P., Kim, H., & Chua, L. O. (2013). Composite behavior of multiple memristor circuits. *IEEE Transactions on Circuits and Systems I: Regular Papers, 60*(10), 2688–2700.

132. Shin, S., Zheng, L., Weickhardt, G., Cho, S., & Kang, S.-M. S. (2013). Compact circuit model and hardware emulation for floating memristor devices. *IEEE Circuits and Systems Magazine, 13*(2), 42–55.

133. Yu, D., Iu, H. H.-C., Liang, Y., Fernando, T., & Chua, L. O. (2015). Dynamic behavior of coupled memristor circuits. *IEEE Transactions on Circuits and Systems I: Regular Papers, 62* (6), 1607–1616.

134. Sozen, H., & Cam, U. (2016). Electronically tunable memristor emulator circuit. *Analog Integrated Circuits and Signal Processing, 89*(3), 655–663.

135. El-Hassan, N. H., Kumar, T. N., & Almurib, H. A. F. (2017). Phase change memory cell emulator circuit design. *Microelectronics Journal, 62*, 65–71.

136. Alharbi, A. G., Fouda, M. E., Khalifa, Z. J., & Chowdhury, M. H. (2016). *Simple generic memristor emulator for voltage-controlled models*. In 59th International Midwest Symposium on Circuits and Systems (MWSCAS), IEEE, pp. 1–4.

137. Alharbi, A. G., Fouda, M. E., Khalifa, Z. J., & Chowdhury, M. H. (2017). Electrical nonlinearity emulation technique for current-controlled Memristive devices. *IEEE Access, 5*, 5399–5409.

138. Data Sheet AD844: Current Feedback Operational Amplifier (CFOA). Online. Available: www.analog.com. Accessed 26 Oct 2017.
139. Data Sheet AD633: Four-quadrant, Analog Multiplier. Online. Available: www.analog.com. Accessed 26 Oct 2017.
140. Yu, D., Zheng, C., Iu, H. H.-C., Fernando, T., & Chua, L. O. (2017). A new circuit for emulating Memristors using inductive coupling. *IEEE Access, 5*, 1284–1295.
141. Senani, R., Bhaskar, D., & Singh, A. (2014). *Current conveyors: Variants, applications and hardware implementations*. Cham: Springer.

Index

© Springer Nature Switzerland AG 2021
A. G. Alharbi, M. H. Chowdhury, *Memristor Emulator Circuits*,
https://doi.org/10.1007/978-3-030-51882-0

Printed in the United States
by Baker & Taylor Publisher Services